T0315042

**Intelligent Satellite Design
and Implementation**

IEEE Press
445 Hoes Lane
Piscataway, NJ 08854

IEEE Press Editorial Board
Sarah Spurgeon, *Editor in Chief*

Jón Atli Benediktsson

Anjan Bose

James Duncan

Amin Moeness

Desineni Subbaram Naidu

Behzad Razavi

Jim Lyke

Hai Li

Brian Johnson

Jeffrey Reed

Diomidis Spinellis

Adam Drobot

Tom Robertazzi

Ahmet Murat Tekalp

Intelligent Satellite Design and Implementation

Jianjun Zhang
China Academy of Space Technology
Beijing, China

Jing Li
Beijing Institute of Technology
Beijing, China

IEEE PRESS

WILEY

Copyright © 2024 by The Institute of Electrical and Electronics Engineers, Inc.
All rights reserved.

Published by John Wiley & Sons, Inc., Hoboken, New Jersey.
Published simultaneously in Canada.

No part of this publication may be reproduced, stored in a retrieval system, or transmitted in any form or by any means, electronic, mechanical, photocopying, recording, scanning, or otherwise, except as permitted under Section 107 or 108 of the 1976 United States Copyright Act, without either the prior written permission of the Publisher, or authorization through payment of the appropriate per-copy fee to the Copyright Clearance Center, Inc., 222 Rosewood Drive, Danvers, MA 01923, (978) 750-8400, fax (978) 750-4470, or on the web at www.copyright.com. Requests to the Publisher for permission should be addressed to the Permissions Department, John Wiley & Sons, Inc., 111 River Street, Hoboken, NJ 07030, (201) 748-6011, fax (201) 748-6008, or online at http://www.wiley.com/go/permission.

Trademarks: Wiley and the Wiley logo are trademarks or registered trademarks of John Wiley & Sons, Inc. and/or its affiliates in the United States and other countries and may not be used without written permission. All other trademarks are the property of their respective owners. John Wiley & Sons, Inc. is not associated with any product or vendor mentioned in this book.

Limit of Liability/Disclaimer of Warranty: While the publisher and author have used their best efforts in preparing this book, they make no representations or warranties with respect to the accuracy or completeness of the contents of this book and specifically disclaim any implied warranties of merchantability or fitness for a particular purpose. No warranty may be created or extended by sales representatives or written sales materials. The advice and strategies contained herein may not be suitable for your situation. You should consult with a professional where appropriate. Further, readers should be aware that websites listed in this work may have changed or disappeared between when this work was written and when it is read. Neither the publisher nor authors shall be liable for any loss of profit or any other commercial damages, including but not limited to special, incidental, consequential, or other damages.

For general information on our other products and services or for technical support, please contact our Customer Care Department within the United States at (800) 762-2974, outside the United States at (317) 572-3993 or fax (317) 572-4002.

Wiley also publishes its books in a variety of electronic formats. Some content that appears in print may not be available in electronic formats. For more information about Wiley products, visit our web site at www.wiley.com.

Library of Congress Cataloging-in-Publication Data
Names: Zhang, Jianjun (Writer on artificial satellites), author. | Li,
Jing, 1982- author.
Title: Intelligent satellite design and implementation / Jianjun Zhang,
    Jing Li.
Description: Hoboken, New Jersey: Wiley, [2024] | Includes index.
Identifiers: LCCN 2023028789 (print) | LCCN 2023028790 (ebook) | ISBN
    9781394198955 (hardback) | ISBN 9781394198962 (adobe pdf) | ISBN
    9781394198979 (epub)
Subjects: LCSH: Artificial satellites. | Artificial intelligence.
Classification: LCC TL796 .Z429 2024 (print) | LCC TL796 (ebook) | DDC
    629.460285/63–dc23/eng/20230909
LC record available at https://lccn.loc.gov/2023028789
LC ebook record available at https://lccn.loc.gov/2023028790

Cover Design: Wiley
Cover Image: © Adastra/Getty Images

Set in 9.5/12.5pt STIXTwoText by Straive, Pondicherry, India

Contents

About the Authors

Jianjun Zhang, PhD, Professor

He received PhD degree from the Institute of Optoelectronics, Chinese Academy of Sciences, in 2010. He is a professor at the Beijing Institute of Spacecraft System Engineering, China Academy of Space Technology. He is also member of the Youth Science Club of China Electronics Society, member of the Edge Computing Expert of China Electronics Society, Chairman of the "Space (Aerospace) Information Technology" Professional Committee of China Electronics Society, and member of the Satellite Application Expert Group of China Aerospace Society. He is chiefly engaged in satellite navigation system design and advanced spatial information system technology based on cognitive mechanism. He has presided over several major projects such as the National Natural Science Foundation's major research project, the final assembly fund, the 863 project, and the development project of the Science and Technology Commission of the China Academy of Space Technology. He has published more than 50 SCI/EI search papers in international journals and conferences, authorized more than 20 invention patents at home and abroad, and published 3 monographs. He won the third prize of National Defense Science and Technology Progress Award.

Jing Li, PhD, Associate Professor, Supervisor

She received PhD degree from Beijing Institute of Technology in 2011. She is an associate professor of School of Automation, Beijing Institute of Technology. She is an expert member of the "Space Information Technology" Youth Committee of China Electronics Society; her main research direction is robot environmental awareness, image detection and target tracking, and multi-sensor information fusion. She has presided over more than 10 projects including the National Natural Science Youth Fund, the Postdoctoral Special Fund, the Key Laboratory of the Ministry of Education, and the Science and Technology Cooperation. She has published 25 academic papers (including 10 SCI papers and 15 EI papers) and the book "Image Detection and Target Tracking Technology" and has been granted

7 national invention patents as the first author and the National Science and Technology Progress Award 2 (ranked 8th). The postgraduates whom she guided have been awarded the second prize of the 14th China Graduate Electronic Design Competition, the second prize of the first China-Russia (Industrial) Innovation Competition, and the second prize of the 14th National College Student Smart Car Competition.

Preface

Artificial Intelligence (AI) is an interdisciplinary subject developed by integrating computer science, cybernetics, information theory, neurophysiology, psychology, linguistics, philosophy, and other disciplines. It is one of the three cutting-edge technologies in the 21st century (genetic engineering, nanoscience, AI). Through AI technology, machines can be competent for some complex tasks that usually require human intelligence to complete, thus greatly simplifying manual operations, improving production efficiency, and improving production relations. It is a very disruptive, cutting-edge technology.

At present, the international space power led by NASA has taken space as an important stage for AI to play its role. Many space tasks that have been carried out or will be carried out have more or less adopted AI technology to improve the efficiency of related tasks. Although the current application of AI technology in space is still limited and the achievements are not outstanding enough, the power of AI has been demonstrated, and the future development direction it represents has also begun to emerge. The successful application of AI in various fields has laid a good foundation for the design of satellite systems and the development of satellite intelligence in the future.

In a series of processes such as satellite development, testing, flight control, delivery and use, the problems of the unattended space environment, the high cost of testing and maintenance, and many factors of fault problems have been puzzling scientific researchers. AI-supporting satellite system technology is a powerful means to solve these problems and is one of the development directions of satellite system design in the future. In the future, it will not only be able to process complete information but also process incomplete information, and even intelligently supplement incomplete information, and make the processing of information and data more mature, efficient, and accurate according to the feedback system. At the same time, experience is constantly accumulated in daily operation, so that the AI system can adapt to the changing environment, gradually realize the automatic evolution mechanism, and make the AI system itself constantly learn, changing the single passive processing information into active, intelligent processing information, and even have a certain predictive ability.

With the increasing development of AI algorithms and application technologies, the next development of intelligent satellites will focus on all aspects: developing the design of onboard intelligent chips to lay the hardware foundation for satellite intelligence. Develop satellite system design based on AI to realize a processing platform that can meet the flexible expansion of multiple tasks and support the flexible reconfiguration of system resources in case of failure. Develop the on-orbit fault detection and maintenance technology based on AI to realize the monitoring of satellite on-orbit status. Carry out research on satellite intelligent control technology based on AI, and realize the application of real-time intelligent autonomous attitude control, intelligent autonomous GNC, and intelligent information technology in aerospace control systems, platforms, and payloads. Carry out research on satellite-ground integration technology based on AI and build a satellite-ground integration satellite platform. Finally, combined with intelligent learning algorithm, the intelligent task of satellite platform is realized.

Jianjun Zhang
China Academy of Space Technology
Beijing, China

Jing Li
Beijing Institute of Technology
Beijing, China

1

Development of Artificial Intelligence

1.1 The Concept and Evolution of Artificial Intelligence

1.1.1 The Concept of Artificial Intelligence

Artificial intelligence (AI), also known as machine intelligence, refers to the intelligence represented by machines made by people. Generally, AI refers to human intelligence technology realized by means of various ordinary computer programs. The definition in the general textbook is "the research and design of intelligent agent," which refers to a system that can observe the surrounding environment and make actions to achieve goals [1, 2].

The definition of AI can be divided into two parts, namely, "artificial" and "intelligence." "Artificial" is easier to understand and less controversial. Sometimes we consider what humans can do and create, or whether a person's own level of intelligence is high enough to create AI, and so on. But to sum up an "artificial system" is an artificial system in the general sense. There are many questions about what "intelligence" is. This involves other issues such as consciousness, self and mind, including the unconscious mind. The only intelligence that people know is their own intelligence, which is a widely accepted view. But our understanding of our own intelligence is very limited, and our understanding of the necessary elements of human intelligence is also very limited, so it is difficult to define what "intelligence" is made by "artificial." Therefore, the research on AI often involves the research on human intelligence itself. Other intelligence with animals or other artificial systems is also generally considered as a research topic related to AI.

A popular definition of AI, as well as an earlier definition in this field, was put forward by John McCarthy of the Massachusetts Institute of Technology at the Dartmouth Conference in 1956: AI is to make the behavior of machines look like

Intelligent Satellite Design and Implementation, First Edition. Jianjun Zhang and Jing Li.
© 2024 The Institute of Electrical and Electronics Engineers, Inc. Published 2024 by John Wiley & Sons, Inc.

that of human beings. But this definition seems to ignore the possibility of strong AI. Another definition is that AI is the intelligence represented by artificial machines. In general, the current definition of AI can be divided into four categories, namely, machines "think like people," "move like people," "think rationally," and "act rationally." Here, "action" should be broadly understood as the decision to take action or specify action, rather than physical action.

Strong AI believes that it is possible to produce intelligent machines that can really reason and solve problems, and such machines will be considered as perceptual and self-conscious. There are two types of strong AI:

1) Human-like AI, that is, the thinking and reasoning of machines, is like human thinking.
2) Nonhuman AI, that is, machines produce perception and consciousness completely different from human beings and use reasoning methods completely different from human beings.

The term "strong artificial intelligence" was originally created by John Rogers Hiller for computers and other information-processing machines. Its definition is: strong AI holds that computers are not only a tool for studying human thinking. On the contrary, as long as it runs properly, the computer itself is thinking. The debate on strong AI is different from the debate on monism and dualism in a broader sense. The main point of the argument is: if the only working principle of a machine is to convert encoded data, then is the machine thinking? Hiller thought it was impossible. He gave an example of a Chinese room to illustrate that if the machine only converts data, and the data itself is a coding representation of some things, then without understanding the correspondence between this coding and the actual things, the machine cannot have any understanding of the data it processes. Based on this argument, Hiller believes that even if a machine passes the Turing test, it does not necessarily mean it is really thinking and conscious like a person. There are also philosophers who hold different views. Daniel Dennett believes in his book *Consciousness Explained* that man is just a machine with a soul. Why do we think: "Man can have intelligence, but ordinary machines can't?" He believes that it is possible to have thinking and consciousness when data is transferred to machines like the above.

The weak AI point of view believes that it is impossible to produce intelligent machines that can really reason and solve problems. These machines just look intelligent, but they do not really have intelligence, nor do they have autonomous consciousness. Weak AI came into being when compared with strong AI because the research on AI was at a standstill for a time, and it began to change and go far ahead until the artificial neural network (ANN) had a strong computing ability to simulate. In terms of the current research field of AI, researchers have created a large number of machines that look like intelligence, and obtained quite fruitful theoretical and substantive results, such as the Eureqa computer program developed by Cornell University Professor Hod Lipson and his doctoral student Michael Schmidt in 2009, as long as some data are given, This computer program can deduce the Newtonian mechanics formula that Newton spent years of research to discover in only a few hours, which is equivalent to rediscovering the Newtonian mechanics formula in only a few hours. This computer program can also be used to study many scientific problems in other fields. These so-called weak AI have made great progress under the development of neural networks, but there is no clear conclusion on how to integrate them into strong AI.

Some philosophers believe that if weak AI is realizable, then strong AI is also realizable. For example, Simon Blackburn said in his introductory philosophy textbook Think that a person's seemingly "intelligent" action does not really mean that the person is really intelligent. I can never know whether another person is really intelligent like me or whether they just look intelligent. Based on this argument, since weak AI believes that it can make the machine look intelligent, it cannot completely deny that the machine is really intelligent. Blackburn believes that this is a subjective issue. It should be pointed out that weak AI is not completely opposite to strong AI, that is, even if strong AI is possible, weak AI is still meaningful. At least, the things that computers can do today, such as arithmetic operations, were considered to be in great need of intelligence more than 100 years ago. And even if strong AI is proven to be possible, it does not mean that strong AI will be developed.

In a word, AI is an interdisciplinary subject, belonging to the intersection of natural science and social science. The disciplines involved include physics, philosophy and cognitive science, logic, mathematics and statistics, psychology, computer science, cybernetics, determinism, uncertainty principle, sociology, criminology, and intelligent crime. The research on AI is highly technical and professional, and each branch field is deep and different, so it covers a wide range. The core issues of AI include the ability to construct reasoning, knowledge, planning, learning, communication, perception, movement, and operation of objects that are similar to or even beyond human beings. Strong AI is still the long-term goal in this field.

There are different understandings about the definition of AI from different perspectives, and there is no unified definition at present. Here are some definitions with high acceptance:

1) In 1956, scientists such as John McCarthy of Stanford University, Marvin Minsky of Massachusetts Institute of Technology, Herbert Simon and Allen Newell of Carnegie Mellon University, Claude Shannon of Bell Laboratory, and other scientists first established the concept of "artificial intelligence" in Dartmouth College of the United States, that is to say, let machines recognize, think, and learn like people, that is, use computers to simulate human learning

and other aspects of intelligence. At the same time, seven typical task directions have been established: machine theorem proving, machine translation, expert system, game, pattern recognition, learning, robot, and intelligent control.

2) In 1981, A. Barr and E.A. Feigenbaum proposed from the perspective of computer science: "Artificial intelligence is a branch of computer science. It is concerned with the design of intelligent computer systems, which have intelligent characteristics associated with human behavior, such as understanding language, learning, reasoning, problem solving, etc."

3) In 1983, Elaine Rich proposed that "Artificial intelligence is to study how to use computers to simulate human brain to engage in reasoning, planning, design, thinking, learning and other thinking activities, and to solve complex problems that are still considered to need to be handled by experts."

4) In 1987, Michael R. Genesereth and Nils J. Nilsson pointed out that "AI is the science of studying intelligent behavior, and its ultimate purpose is to establish a theory on the behavior of natural intelligence entities and guide the creation of artificial products with intelligent behavior."

5) Wikipedia defines "AI is the intelligence displayed by machines," that is, as long as a certain machine has some or some "intelligence" characteristics or performance, it should be counted as "AI." The Encyclopedia Britannica defines "AI is the ability of a digital computer or a robot controlled by a digital computer to perform some tasks that intelligent organisms only have." Baidu Encyclopedia defines AI as "a new technological science that studies and develops theories, methods, technologies, and application systems for simulating, extending and extending human intelligence," regards it as a branch of computer science, and points out that its research includes robots, language recognition, image recognition, natural language processing, and expert systems.

6) According to the White Paper on the Standardization of Artificial Intelligence (2018), "Artificial intelligence is a theory, method, technology and application system that uses digital computers or machines controlled by digital computers to simulate, extend and expand human intelligence, perceive the environment, acquire knowledge and use knowledge to obtain the best results. Artificial intelligence is the engineering of knowledge, which is the process that machines imitate human beings to use knowledge to complete certain behaviors."

7) The US Defense Authorization Act of FY 2019 defines AI as all systems that can "act rationally" in a changing and unpredictable environment without sufficient human supervision or can learn from experience and use data to improve performance.

To sum up, AI allows computer/machine to simulate, extend, and expand human intelligence, so that the system can cope with changing and unpredictable environments without human supervision, deal with complex problems through learning and experience, and achieve performance improvement [2, 3].

1.1.2 Evolution of Artificial Intelligence

AI was born in the 1950s. Over the past 60 years, it has experienced three waves (Figure 1.1) and has formed such theoretical schools as semiotics, connectionism, and behaviorism. The first wave (1956–1976) was dominated by semiotics, and its main achievements were machine reasoning, expert system, and knowledge engineering; the second wave (1976–2006), dominated by the connectionist school, produced neural networks, machine learning, and other achievements; the third wave (from 2006 to now), led by the connecting school, achieved great success in deep neural network and deep reinforcement learning, and AI entered the stage of commercial development. The behavioral school theory has not played a leading role in the third wave so far. Its core is adaptive control, learning control, evolutionary computing, and distributed intelligence, which are the basis of modern control theory [3].

In the three waves of AI development, the key points in technology and application breakthrough are shown in Figure 1.2.

Notably, the behavioral school theory has not played a leading role in the third wave so far. Its core is adaptive control, learning control, evolutionary computing, and distributed intelligence, which is the basis of modern control theory. In recent years, different voices have emerged in the development of AI. They believe that AI technology based on the connectionist school theory has a huge gap in solving problems that cannot be reasoned and counted and cannot achieve strong AI with creative thinking. They believe that the symbolic school and behavioral school theory pay more attention to the abstraction of human higher intelligence and

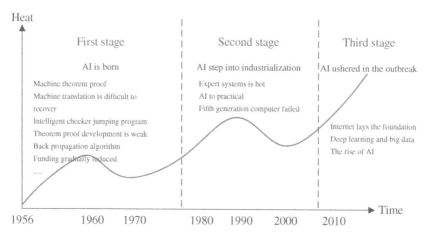

Figure 1.1 Three waves of AI.

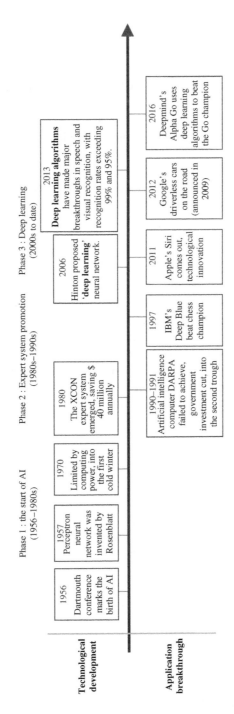

Figure 1.2 Key points of AI technology development.

more attention to adaptation and evolution; the realization of strong AI in the future may need to find another way.

1.2 The Current Scope and Technical Framework of Artificial Intelligence

1.2.1 Technical Scope

Compared with the early AI, the new generation AI is under the guidance of the new information environment, massive database, and continuously evolving and enriching strategic objectives, relying on the two basic platforms of cloud computing and big data and the three general technologies of machine learning, pattern recognition, and human–computer interaction and taking the new computing architecture, general AI, and open source ecosystem as the main guidance, and continues to build and improve the technical framework system. As a result of multi-disciplinary intersections and universal technology, AI technology has formed a complex network of technology systems together with related downstream technologies and applications. At present, the network is in its infancy, but it is still in a dynamic state of rapid renewal and drastic changes [4].

1.2.2 Technical Framework

According to the description in the White Paper on the Development of New Generation AI (2017), the current AI technical framework is mainly composed of three parts: basic layer, technical layer, and application layer, as shown in Figure 1.3:

1) Basic layer
   The basic layer mainly includes big data, smart sensors, smart chips, and algorithm models. Among them, smart sensors and smart chips belong to basic hardware, and algorithm models belong to core software [5].
2) Technical level
   The technical level mainly includes pattern recognition, autonomous planning, intelligent decision-making, autonomous control, human–machine cooperation, group intelligence, etc.
3) Application layer
   The application layer mainly includes intelligent agents (robots, unmanned driving, intelligent search, and unmanned aerial vehicles) and industrial applications (finance, medical, security, education, human settlements, etc.).

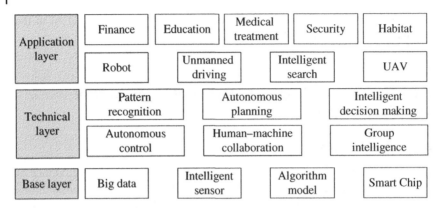

Figure 1.3 General technical system of AI technology.

1.2.3 Technical Features

At present, the development of AI is based on the theory of connectivity school. The main driving factors are explosive growth of data, continuous improvement of computing power, continuous optimization of algorithm models, and deep coupling of capital and technology. The main technologies include computer vision, machine learning, natural language processing, robot technology, speech recognition, etc. At present, intelligent medical, finance, education, transportation, security, home manufacturing, and other fields have been widely used and developed rapidly in unmanned driving, intelligent robots, and others. Among them, deep learning is a typical representative of the development of the new generation of AI, but "deep learning" is not a synonym for AI [6, 7].

Driven by data, computing power, algorithm model, and multiple applications, AI is evolving from auxiliary equipment and tools to assist and partner for collaborative interaction and is more closely integrated into human production and life (Figure 1.4):

1) Big data has become the cornerstone of the sustained and rapid development of AI.
2) Text, image, voice, and other information realize cross-media interaction.
3) Network-based swarm intelligence technology has begun to sprout.
4) Autonomous intelligent system has become an emerging development direction.
5) Human–machine collaboration is giving birth to a new type of hybrid intelligence.

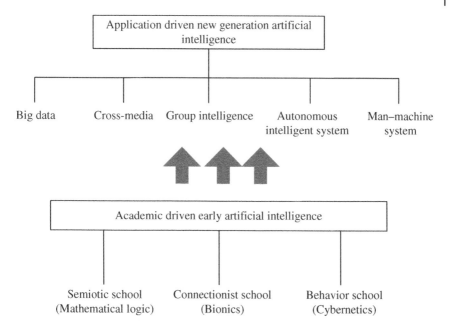

Figure 1.4 Technical characteristics of new generation AI technology.

1.3 The Overall Development Trend of Artificial Intelligence

1.3.1 Current Development Trend

1) **Artificial intelligence technology, as an important factor leading industry disruption and technological change, has achieved explosive growth in all fields around the world.**

   In the past decade, intelligent algorithms represented by deep learning and intensive learning have promoted AI to continue to exert its power in cutting-edge technologies and applications, and giant companies such as Google and IBM have made major breakthroughs in search engines, human–computer games, intelligent medical care, unmanned driving, and other fields. By December 2018, more than 5000 AI startups had been born and flourished in reading and writing assistant, financial industry, enterprise management, advanced manufacturing, and other industries, which had a profound impact on various fields [8–10].

2) **The major developed countries in the world regard the development of AI as a major strategy to enhance national competitiveness and maintain national security.**

The United States was the first country to release a national strategy for AI. In 2016, it released the National Strategic Plan for AI Research and Development. In the next two years, a total of 15 countries and regions released their AI strategies and accelerated their planning and layout in the field of AI (Figure 1.5). Among them, thanks to decades of federal research funds, industrial production, academic research, and the inflow of foreign talents, the United States has been leading the global wave of AI development in basic theory, software and hardware, talents, enterprises, and other aspects.

3) **In many aspects, AI is far from mature application, and there are still technical bottlenecks.**

At present, it is generally believed that AI has surpassed human beings in large-scale image recognition. Although some progress has been made in machine translation, it is not close to the ideal level, while there is a big gap in chat conversation. In terms of driverless driving, the current commercial autonomous vehicle is auxiliary driving, and the real autonomous vehicle is still in the development and testing stage, in which dealing with emergency and abnormal traffic conditions is the difficulty encountered by AI. Similarly, in speech recognition, although AI recognition is close to humans in the experimental environment, in reality, especially in the presence of environmental interference, the recognition rate of AI is actually unsatisfactory, and it often makes some common sense mistakes that humans cannot make. Some experts even believe that AI technology will not make a major breakthrough in 20 years, because there are not many new topics in the field of AI research done by the current academic community. Even if scientists work hard, it will take about 20 years to accumulate a theoretical basis that makes people feel very excited and surprised.

4) **AI will accelerate the intelligent transformation and upgrading of traditional industries.**

In the future, with the continuous improvement of AI capabilities, the integration of AI and traditional industries will continue to deepen, driving the transformation and upgrading of traditional industries to intelligence and at the same time creating huge benefits for society. According to the research report, by 2025, the market scale of AI in agricultural applications will reach 20 billion dollars, and the financial industry will save and generate 34–43 billion dollars in costs and new market opportunities, and the annual expenditure in the medical field will be reduced by 54 billion dollars.

5) **Human–computer cooperation will become the main direction of AI commercialization.**

With the complementary nature between humans and AI systems, the collaborative interaction between humans and AI systems will become the main direction of AI commercialization. Although fully autonomous AI systems

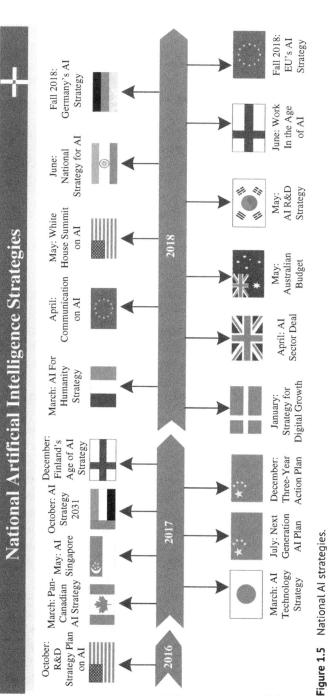

Figure 1.5 National AI strategies.

will play an important role in underwater or deep space exploration and other applications, AI systems cannot completely replace humans in the short term in disaster recovery, medical diagnosis, and other applications.

Human–machine cooperation can be divided into three ways, namely, joint execution, auxiliary execution, and alternative execution:

- Joint execution

  AI system positioning: perform peripheral tasks supporting human decision makers.

  Typical applications: short-term or long-term memory retrieval and prediction tasks.

- Auxiliary execution

  AI system positioning: when human beings need help, the artificial intelligence system performs complex monitoring functions.

  Typical applications: ground proximity alarm system, decision-making, and automatic medical diagnosis in aircraft.

- Alternative execution

  AI system positioning: AI systems perform tasks with very limited capabilities for humans.

  Typical applications: complex mathematical operations, dynamic system control and guidance in controversial operating environment, automatic system control in hazardous or toxic environment, nuclear reactor control room, and other rapid response systems.

6) **AI will present a competitive pattern of leading platform plus scenario application.**

Under the trend of AI platformization, AI will present a competitive pattern of several leading platforms and extensive scene applications in the future, and ecological builders will become the most important model among them. As shown in Figure 1.6, the future AI competition pattern will mainly present five modes.

Mode 1: Ecological construction – take the whole industrial chain ecology and scenario application as a breakthrough. Take Internet companies as the main body, mainly invest in infrastructure and technology for a long time. At the same time, scenario applications will be used as traffic entry, accumulate applications, and become the leading application platform. It will become the builder of AI ecosystem.

Mode 2: Technical algorithm-driven – take technical layer and scene application as a breakthrough. Take software companies as the main body, deeply cultivate algorithm platform and general technology platform, and gradually establish application platform with scene application as the traffic entrance.

Mode 3: Application focus – scene application. Based on the scenario or industry data, many segmented scenario applications are developed mainly for startups and traditional industry companies.

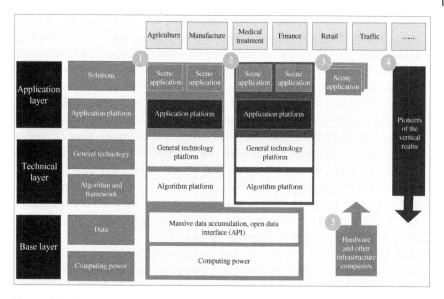

Figure 1.6 Competition pattern of AI in the future.

Mode 4: Vertical field first – killer application and gradually builds vertical field ecology. It is mainly a pioneer in the vertical field. It relies on killer applications to accumulate a large number of users and data in the vertical field, and deeply cultivates the common technologies and algorithms in the field, becoming a subverter in the vertical field.

Mode 5: Infrastructure provision – starting from infrastructure and expanding to the downstream of the industrial chain. Take chip or hardware and other infrastructure companies as the main part, start from infrastructure, improve technical capabilities, and expand to the upstream of data, algorithms, and other industrial chains.

1.3.2 Technical Development Trend

At present, AI, represented by deep learning and big data, has made amazing progress in the fields of image classification, speech recognition, visual understanding, and machine translation. However, deep learning relies on annotated data and lacks the ability to express logical reasoning and causal relationships. It is difficult to deal with tasks with complex spatiotemporal correlations and cannot achieve the goal of strong AI. In the future, AI technology needs to continue to break through [10–12].

1) *From deep learning to brain intelligence algorithm*

   In the aspect of algorithms, the basis of AI technology is developed according to two main lines: deep learning improvement and new algorithms. Advanced in-depth learning focuses on breaking through the methods of adaptive learning, autonomous learning, zero-data learning, unsupervised learning, migration learning, etc., to achieve AI with high reliability, high interpretability, and strong generalization ability. Second, the academic community continues to explore new algorithms, and brain-like intelligent algorithms have become a frontier hotspot, focusing on breaking through brain-like information coding, processing, memory, learning, association, reasoning, and other theories, forming brain-like complex systems and brain-like control and other theories and methods, and establishing new models of large-scale brain-like intelligent computing and brain-inspired cognitive computing models [12, 13].

2) *From dedicated intelligence to general intelligence*

   With the continuous development of science and technology and the deep transformation of social structure, the problems facing human beings are highly complex. There is an urgent need for a wide range, high integration, and strong adaptability of general wisdom in professional fields such as game, identification, control, and prediction, which significantly improves the ability of human beings to read, manage, and reorganize knowledge. General AI has the characteristics of reducing dependence on domain knowledge, improving the applicability of processing tasks, and achieving the correction of machine-independent cognition. It has the ability to process multiple types of tasks and adapt to unexpected situations. Its substantive progress will truly start the prelude of intelligent revolution, highly integrate with the existing physical and information world, and profoundly affect all aspects of social and economic development.

3) *From machine intelligence to human–machine hybrid intelligence*

   Machine intelligence (AI) and human intelligence have their respective strengths and need to learn from each other and complement each other. Integrating multiple intelligence models, human–machine coexistence will become the new normal of the future society. It will focus on breaking through the theory of human–machine win–win situation understanding and decision learning, intuitive reasoning and causal model, memory and knowledge evolution, and realize hybrid enhanced intelligence that learning and thinking are close to or exceed the level of human intelligence.

4) *From single agent intelligence to group intelligence*

   Group intelligence originates from the research on the group behavior (including information transmission, collective decision-making, etc.) of social insects represented by ant colonies, bee colonies, etc. The group has self-organization, division of labor, and coordination. Group intelligence refers to the collective intelligent behavior of a system composed of individual units through interaction between each other or with the environment, with the characteristics of intelligent emergence. Science believes that the era of group intelligence based on networks is coming. Group intelligence will focus on breaking through the theories and methods of organization, emergence and learning of group intelligence, establishing expressive and computable group intelligence incentive algorithms and models, and forming group intelligence capabilities based on a network environment [14, 15].

5) *Weak AI will gradually transform into strong AI*

AI can be divided into three stages: weak AI, strong AI, and super AI. The level of AI in the three stages has been continuously improved. Weak AI refers to human ability in some fields; strong AI refers to having human capabilities in all fields, being able to compete with human beings in all aspects, and being unable to simply distinguish between human beings and machines; super AI means that it can surpass human beings in all fields, and can surpass human beings in the fields of innovation and creative creation to solve any problems that cannot be solved by human beings.

From the current development level of AI, AI is still a weak AI based on specific application fields, such as image recognition, speech recognition, and other biometric analysis, such as intelligent search, intelligent recommendation, intelligent sorting, and other intelligent algorithms. When it comes to vertical industries, AI mostly assists human beings in their work in the role of assistant, such as the current intelligent investment advisor and autonomous vehicle, but the AI that completely gets rid of human beings and can reach or even surpass human beings in the true sense cannot be realized. In the future, with the substantial growth of computing power and data volume and the improvement of algorithms, weak AI will gradually transform into strong AI, and machine intelligence will advance from perception, memory, and storage to cognition, autonomous learning, decision-making, and execution.

6) *Data, algorithm, and computing power are the three carriages of AI*

In recent years, with the increase of the data volume index, the algorithm theory continues to be updated iteratively, and the computing ability continues to be enhanced, which jointly promotes the accelerated development of AI. The relationship between the three and AI is shown in Figure 1.7.

With the rapid development of the Internet, not only the standard training set and test set data are increasing but also the massive and high-quality application scenario data are becoming increasingly rich. Real-time, massive, multi-sources, and multi-types of data can describe reality more closely from different angles, and machine learning algorithms can be used to mine multi-level associations between data, laying the foundation of data sources for AI applications.

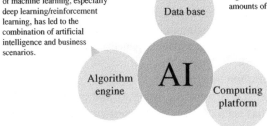

Machine learning algorithm is the engine for realizing artificial intelligence landing
The improvement and iteration of machine learning, especially deep learning/reinforcement learning, has led to the combination of artificial intelligence and business scenarios.

The large amount of real -time data generated lays the foundation for the application of artificial intelligence.
Training artificial intelligence algorithm model through large amounts of data.

Deep learning has higher requirements for parallel computing and data throughput per unit time.
With the development of GPU/FPGA and the improvement of computing power, the cloud computing platform can quickly calculate and process large amounts of data.

Figure 1.7 Relationship between data, algorithm, calculation, and artificial intelligence.

The development of algorithms, especially the paper published by Professor Geoffrey Hinton in 2006, has started the wave of deep learning in academia and industry. The deep learning algorithm represented by ANN has become the core engine of AI applications.

AI has a high demand for computing power. With the rapid development of cloud computing technology and chip processing power, thousands of machines can be used for parallel computing. In particular, the development of tensor processing unit (TPU), graphics processing unit (GPU), field programmable gate array (FPGA), and special chips for AI has laid the foundation for implementing AI computing power, making the application of AI similar to human's deep neural network algorithm model a reality.

1.4 The Main Achievements of AI

1.4.1 Image Recognition

Image recognition is more troublesome than speech recognition because the object of speech recognition is always a variety of limited languages. However, the recognition of the face and cat in the image cannot be handled by a single model when it comes to the specific implementation. The current state is that if you select a point, such as face recognition, and hit dozens of PhDs and hundreds of GPUs, it will take one to two years. If you can still find the landing point and continue to obtain data, you can achieve very high accuracy (more than 99%). But this accuracy cannot cover other fields at once. You can only do it point by point. Because there are practical landing scenarios (banks, etc.) in such fields as face, they have developed rapidly. It will take some time for other fields to reach the same level.

In 2009, a Chinese scholar in the Department of Computer Science at Princeton University (the first author is Jia Deng) published the paper "ImageNet: A large scale hierarchical image database," announcing the establishment of the first super-large image database for use by computer vision researchers. In 2010, ImageNet Large Scale Visual Recognition Challenge 2010 (ILSVRC2010), a large-scale image recognition contest based on ImageNet, was held for the first time. The initial rule of the competition is to take 1.2 million images in the database as training samples, which belong to more than 1000 different categories and are manually marked. The trained program will be used to evaluate 50000 test images to see whether it is accurate in image classification.

In 2012, Professor Hinton and his two graduate students, Alex Krizhevsky and Illya Sutkerver, applied the latest technology of deep learning to the problem of ImageNet. Their model is an eight-layer convolutional neural network with 650000 neurons and 60 million free parameters. This neural network uses the discard algorithm and the excitation function of the rectified linear unit (RELU) introduced in the previous two articles. Professor Hinton's team used two GPUs to train the program with 1.2 million images, which took nearly six days. The trained model faces 150000 test images, and the error rate of the first five categories predicted is only 15.3%. In the 2012 ImageNet competition with 30 groups, the test result ranked first. The second is from the model of the Japanese team, and the corresponding error rate is as high as 26.2%. This marks that neural network has greatly surpassed other technologies in the field of image recognition and has become a turning point in the breakthrough of AI technology.

In the ImageNet image recognition competition in December 2015, the team from Microsoft Research Asia (MSRA) won the championship. As the depth of the network increases, the efficiency of learning decreases. In order to solve the problem of attenuation of effective information in layer-by-layer transmission, the MSRA team tried an algorithm called "deep residual learning." MSRA hopes that in the deep residual learning model, using the 152 layers of neural network, the recognition error rate of the first five categories has reached a new low of 3.57%, which is lower than about 5% of the error rate of a normal person.

1.4.2 Speech Recognition

Speech recognition is a field that has been conquered by deep learning in recent years. If enough money is spent, the recognition accuracy can reach 99%. In speech recognition, improving the accuracy of the last few points is likely to be more difficult than the previous 90% accuracy. However, the accuracy of the last few points often formally crosses the key of being able to use and not being able to use. Before in-depth learning, people have tried to conquer speech recognition for many years. Generally speaking, the first speech recognition system that can recognize the pronunciation of 10 English numbers, studied by Bell Labs in 1952, is considered as the starting point of speech recognition. In this way, human beings have worked hard on this matter for more than 60 years. In the 1970s, speech recognition with a small vocabulary was completed, and in the 1980s, speech recognition with a large vocabulary was completed. Then the accuracy was stuck there, hovering around 85%, and one card was close to 30 years. Microsoft and IBM have tried to apply this technology in the past, but obviously, it has no effect. Many people do not even remember what they did. After the introduction of deep learning into speech recognition, things have changed fundamentally. Now, as long as there is enough data for training, most companies can train enough accurate speech recognition models themselves. This technology basically needs to be commercialized, and it can be more and more regarded as a technology that can be solved without too much investment.

RNN, also known as recurrent neural network or multilayer feedback neural network, is another very important type of neural network. In essence, the difference between RNN and feedforward network is that it can retain a memory state to process the input of a sequence, which is particularly important for handwritten character recognition, speech recognition, and natural language processing.

In October 2012, Geoffrey Hinton, Deng Li, and several other researchers representing four different institutions (University of Toronto, Microsoft, Google, and IBM) jointly published the paper "Application of Deep Neural Network in Acoustic Model of Speech Recognition: Common Viewpoint of Four Research Groups." The researchers borrowed the algorithm of "restricted Boltzmann machine" used by Hinton to "pre-train" the neural network. The depth neural network model is used to estimate the probability of recognizing characters. In a voice input benchmark test conducted by Google, the word error rate reached 12.3%.

In March 2013, Alex Graves of the University of Toronto led the publication of the paper "Deep Circulation Neural Network for Speech Recognition." The paper uses RNN/long short-term memory (LSTM) technology – a network with three hidden layers and 4.3 million free parameters. In a benchmark test called TIMIT, the "phoneme error rate" reached 17.7%, which is superior to the performance level of all other technologies in the same period.

In May 2015, Google announced that by relying on RNN/LSTM-related technologies, the word error rate of Google Voice decreased to 8% (about 4% for normal people).

In December 2015, Dario Amodei of Baidu AI Lab led the paper "End-to-End Speech Recognition in English and Chinese." The model in this paper uses a simplified variant of LSTM, called "closed cycle unit." Baidu's English speech recognition system has received nearly 12 000 hours of speech training. It takes three to five days to complete the training on 16 GPUs. In a benchmark test called WSJ Eval '92, the word error rate was as low as 3.1%, which exceeded the recognition ability of normal people (5%). In another small Chinese test, the recognition error rate of the machine was only 3.7%, while the collective recognition error rate of a five-member group was 4%.

According to this trend, the accuracy of the machine in various benchmark tests of speech recognition will soon catch up with and surpass that of ordinary people. This is another difficulty that AI will overcome after image recognition.

The essence of RNN is that it can handle the output and input (many to many) of a sequence with varying length. In a broad sense, if the traditional feedforward neural network does the optimization of a function (such as image recognition), then what the circular neural network does is the optimization of a program, and the application space is much wider.

Semantic understanding and speech recognition are different from image recognition. Semantic understanding is in a state that is basically unsettled. Many times, when we watch the demonstration, we can see a robot and an intelligent product communicate with people smoothly. There are two possibilities to achieve this. One is cheating, with individuals behind it, which belongs to AI. One is that the conversation is limited to specific scenarios, such as making a phone call in the car or allowing map navigation. The difficulty of semantic understanding is related to the number of concepts to be processed. When the number of concepts to be processed is less than a few thousand, it is still possible to solve the problem in a rule-based way for a specific scenario, which is relatively smooth. But once the scope is extended to the whole social life, it is at most like Google Now and Siri. One of the applications closely related to this is the ability of various intelligent voice assistants in conversation, and the other is translation.

1.4.3 Art Creation

For a long time, people tend to think that machines could understand humans' logical thinking, but they cannot understand humans' rich feelings, let alone humans' aesthetic values. Of course, machines cannot produce works with aesthetic value. Facts speak louder than eloquence. The Alpha Dog took a ground-breaking step against Li Shishi. Mr. Nie Weiping, the chess master, saluted the Alpha Dog's dismounting, which shows that the deep learning algorithm has been able to create aesthetic value. Many chess players spend their entire lives on the chessboard, and what they are looking for is a beautiful and wonderful player. It is so profound, beautiful, mysterious, and abstract that it becomes a dull neuron parameter overnight, which makes many people disillusioned.

In fact, in the field of visual art, the ANN has been able to separate the content and style of a work, at the same time learn the artistic style from the art masters, and transfer the artistic style to another work, rendering the same content with the style of different artists.

This means that the ANN can accurately quantify many vague concepts in the original humanities, such as the "artistic style" in specific fields and the "chess style" in games, and make these skills and styles that can only be understood and cannot be expressed in words simple, easy to copy, and promote.

1.4.4 Other Aspects

In terms of game games, the Deep Q-Network DQN developed by the Google DeepMind team has reached or exceeded the level of human professional players in 29 of the 49 Atari pixel games. Alpha Dog is a top player in Go that completely surpasses humans.

In May 2016, AI Lab from Google reported that the researchers used 2865 English romance novels to train the machine and let the machine learn the narrative and diction style of romance novels. From the perspective of the evolution process of the program, the machine model first comprehends the structure of the space between words and then slowly understands more words. From short to long, the rules of punctuation are slowly mastered, and some sentence structures with more long-term relevance are slowly mastered by the machine.

In May 2016, Google's DeepMind team wrote an article and developed a "Neural Programming Interpreter." This neural network can learn and edit simple programs by itself, which can replace the work of some junior programmers.

APP and mobile phones: Because of the emergence of intelligent photography technology, we have been able to do some new human–computer interactions. In some new fields, whether from the perspective of mobile internet companies, or from the perspective of live broadcasting, online education, mobile games, and other industries, there are several new products to do, representing some new growth in the industry. In the next step, there is new content to be done in the lens module of mobile phones and the industrial chain of mobile phones, from unlocking faces to intelligent photography and other new products.

Security field: At present, face recognition in the security field is changing from single-point camera recognition to the overall recognition of a camera cluster, which is a step in the progress of technology. There are still many technological processes to be completed before we can truly achieve product maturity and user experience maturity. In this process, it needs CPU supercomputing and other related products to drive new markets in smart cities, transportation, government affairs, and other aspects. In general, it is now possible to provide safer and more efficient urban management means and means to catch people in this field in the technical link.

Internet of Things: This is a trillion-level market with many scenarios. Although the needs are different, the technical commonalities are the same: VIP identification, identifying whether you are a VIP member; identity authentication, whether you are yourself; passenger flow analysis, through the data statistics and analysis of a large number of people to better optimize the offline sales, what kind of people to sell, how to lay out the goods and stores, and to carry out optimization and sales promotion.

The maturity of underlying technology drives the emergence of new retail. The corresponding market increment can be found in replacing access control, smart parks, buildings, hospitals, schools, and other scenarios. Accordingly, companies engaged in offline scene integration have relatively potential new market value increments.

Advertisement: The profit from advertising is very high. After the formation of better AI technology, user-generated content (UGC), live broadcast, augmented reality, and other technologies and industrial scale, some new channels have emerged, including the increment of advertising communication, advertising production, advertising services, and other links. In essence, the core is the emergence of new advertising forms; everyone can advertise the products they want to endorse. The live broadcast scenario can also optimize the overall advertising communication efficiency.

Google driverless car: Google Driverless Car is a fully autonomous vehicle developed by Google X Lab of Google, which can start, drive, and stop without the driver. It is currently being tested to drive 480,000 km. The project is led by Sebastian Thrun, the co-inventor of Google Street View. Google engineers used seven test vehicles, six of which were Toyota Prius and one was Audi TT. These cars have been tested on several roads in California, including the Jiuqu Flower Street in the San Francisco Bay Area. These vehicles use cameras, radar sensors, and laser rangefinders to "see" other traffic conditions and use detailed maps to navigate the road ahead. Google said that these vehicles are safer than manned vehicles because they can react more quickly and effectively.

Boston Atlas robot is controlled by a robot. Atlas is a humanoid robot developed by Google's Boston Dynamics. The product iteration of Atlas focused on three major version updates. Atlas robot is based on the early PETMAN humanoid robot of Boston Dynamics, and PETMAN is a humanoid robot designed to detect chemical protective clothing, which can simulate how soldiers act with protective clothing under real conditions. Unlike the limited movement posture of the chemical protective clothing testing machine in the past, PETMAN can not only balance itself and walk freely, bend the body, but also do various calisthenics that have pressure on the chemical protective clothing in the chemical warfare agent operation workshop. PETMAN also simulates the actual test conditions by simulating the human physiology in the protective clothing to control the temperature, humidity, and sweating. Since PETMAN is a humanoid robot used to detect chemical protective clothing, starting from PETMAN, Boston Power Development Department has established Atlas humanoid robot to improve its movement ability. The first version of Atlas robot has four hydraulically driven limbs. Atlas robot is made of aviation grade aluminum and titanium, with a height of about 1.8 m and a weight of 150 kg, and is illuminated by a blue LED. The Atlas robot is equipped with two vision systems – a laser rangefinder and a stereo camera, which are controlled by an onboard computer. Its hands have the ability of fine motor skills. Its limbs have a total of 28 degrees of freedom. The main feature of the first version is that Atlas needs a long cable for power supply. It can walk on a road covered with stones in the laboratory environment and keep its body balanced without falling. Second, in the laboratory environment, let Atlas stand on one foot and introduce external impact. Atlas can still stand on one foot without falling when the impact force is not very large. When walking on the green belt, place obstacles on the path of one foot. When Atlas steps on the obstacles, it will automatically update its gait in the next step and plan the landing point of the next step to leave the obstacles. Finally, carry out the road test in the field environment, and Atlas can also maintain a good balance in the field environment. At the beginning of 2015, in order to participate in the DARPA Robot Challenge at the beginning of June, Atlas completed its own evolution. That is, Atlas has successfully evolved into the second edition. This time, 75% of Atlas's whole body has been redesigned. Only the lower legs and feet have followed the previous version of the design. The new design makes it stronger, faster, and quieter. Thanks to the large battery backpack behind it, it has been freed from the cable. The upgraded Atlas is 1.88 m high and weighs 156.5 kg. The newly improved Atlas is slimmer and smaller than the first version. The more effective airborne hydraulic pump also makes the Atlas robot move faster. The purpose of this improvement is to make Atlas robot easier to complete some challenges. For example, it is required to squeeze into a space specially designed for human beings. So the purpose of this Atlas is to put it in real disaster rescue scenarios. It can communicate with the upper computer, and the protective rope on its head can also be removed. The power supply system of the second version of Atlas is a 3.7-kwh lithium-ion battery pack, which can last for one hour when completing actions including walking, standing, and using tools. The powerful hydraulic pump gives Atlas the ability to stand up when it falls. The third version of Atlas can be operated indoors and outdoors. Similar to the power of the second version, Atlas uses a power supply and hydraulic drive. This also shows that in order to have great power, energy can only be compromised by hydraulic pressure at present. The sensors in Atlas's body and legs keep its body balanced by collecting posture data. The lidar locator and stereo camera on its head can enable Atlas to avoid obstacles, detect ground conditions, and complete cruise tasks. This version of Atlas is 1.75 m high and weighs 82 kg, both of which are lighter than the previous version. In the process of carrying boxes and pushing doors, it is also necessary to mark and complete the object recognition task, and the machine vision ability such as object recognition needs to be improved. The most striking feature of this version is the function that Atlas can climb up automatically after falling down. Compared with other robots, the performance of this function is excellent, and the efficiency of getting up is also very high.

AlphaGo of the game. AlphaGo is the first AI robot to defeat the human professional Go player and the world champion of Go. It is developed by a team led by Demis Hassabis, a Google's DeepMind company. Its main working principle is "deep learning." AlphaGo has used many new technologies, such as neural networks, deep learning, and Monte Carlo tree search, which has made a substantial leap in its strength. AlphaGo is mainly composed of four parts: strategy network, given the current situation, predict, and sample the next move; the goal of fast-moving is the same as that of the strategic network, but the speed is 1000 times faster than that of the strategic network block with an appropriate loss of the quality of moving; value network, given the current situation, it is estimated that the probability of white victory or black victory is high, and Monte Carlo tree search connects the above three parts to form a complete system. The previous version of AlphaGo combines the chess manual of millions of human go experts and the supervised learning of intensive learning for self-training. The capability of AlphaGoZero has been improved qualitatively on this basis. The biggest difference is that it no longer requires human data. In other words, it has not touched human chess at the beginning. The R&D team just let it play chess freely on the chessboard and then conducts a self-game. According to David Silva, the head of the AlphaGo team, AlphaGoZero uses new reinforcement learning methods to make itself a teacher. In the beginning, the system did not even know what is Go, but it started with a single neural network and played it by itself through the powerful search algorithm of the neural network. With the increase of self-game, the neural network is gradually adjusted to improve its ability to predict the next step and finally win the game. What is more, with the deepening of the training, the AlphaGo team found that AlphaGoZero also independently discovered the rules of the game and came up with new strategies, bringing new insights into the ancient game of Go.

References

1 Wang, J.J., Ma, Y.Q., Chen, S.T., et al. (2017). Fragmentation knowledge processing and networked artificial intelligence. *Scientia Sinica (Informationis)*, 47 (2): 171–192.
2 Hudyjaya Siswoyo, J. and Nazim, M.N. (2013). Development of minimalist bipedal walking robot with flexible ankle and split-mass balancing systems. *International Journal of Automation and Computing* 10 (5): 425–437.

3 Zhang, W.X., Ma, L., and Wang, X.D. (2017). Reinforcement learning for event-triggered multi-agent systems. *CAAI Transactions on Intelligent Systems* 12 (1): 82–87.

4 Lake, BM., Salakhutdinov, R., Tenenbaum, J.B. (2015). Human-level concept learning through probabilistic program induction. *Science* 350 (6266): 1332–1338.

5 Gao, X.S., Li, W., and Yuan, C.M. (2013). Intersection theory in differential algebraic geometry: generic intersections and the differential chow form. *Transactions of the American Mathematical Society* 365 (9): 4575–4632.

6 Luo, J.M. (2005). Intelligence and an intelligent model. In: *Proceedings of the Fourth International Conference on Machine Learning and Cybernetics*, 5624–5629. Guangzhou: IEEE Press.

7 Luzeaux, D. and Dalgalarrondo, A. (2001). HARPIC, an hybrid architecture based on representations, percept ion and intelligent control: a way to provide autonomy to robots. In: *Proceedings of the International Conference on Computational Science*, 327–336. London: ACM Press.

8 Hintong, G., Deng, L., Yu, D. et al. (2012). Deep neural networks for acoustic modeling in speech recognition: The shared views of four research groups. *IEEE Signal Processing Magazine* 29 (6): 82–97.

9 Deng, J., Dong, W., Socher, R. et al. (2009). Image Net: A large-scale hierarchical image database. In: *Computer Vision and Pattern Recognition, 2009. CVPR 2009. IEEE Conference on Miami*, 248–255. IEEE.

10 Dan, C.C., Meier, U., Gambardella, L.M. et al. (2010). Deep big simple neural nets excel on handwritten digit recognition. *Neural Computation* 22 (12): 3207–3220.

11 Hinton, G.E., Srivastava, N., Krizhevsky, A. et al. (2012). Improving neural networks by preventing co-adaptation of feature detectors. *Computer Science* 3 (4): 212–223.

12 Glorot, X., Bordes, A., and Bengio, Y. (2011). Deep sparse rectifier neural networks. *Journal of Machine Learning Research* 15: 315–323.

13 Hinton, G.E., Osindero, S., and Teh, Y.W. (2006). A fast learning algorithm for deep belief nets. *Neural Computation* 18 (7): 1527–1554.

14 Zhong, Y.X. (2006). A preliminary investigation on information knowledge intelligence transforms. *Frontiers of Electrical and Electronic Engineering in China* 1 (3): 60–66.

15 Socher, R., Lin, C., Ng, A. Parsing natural scenes and natural language with recursive neural Networks. *Proceedings of the 28th International Conference on Machine Learning (ICML-11)*, Bellevue, WA, USA. Germany (28 June–2 July 2011): International Machine Learning Society, 2011.

2

Artificial Intelligence in the Satellite Field

2.1 The Concept and Connotation of Intelligent Satellite

2.1.1 The Concept of Intelligent Satellite

Satellite system is an engineering system composed of on-orbit operating satellites, application service systems, etc., and completes specific space missions. It is a typical complex large system [1–3].

2.1.1.1 Artificial Intelligence in Satellite Field

Artificial intelligence in satellite field refers to the "special artificial intelligence" that integrates the relevant theories, technologies and methods of artificial intelligence at all levels of physical laws, technologies, and knowledge in satellite field and with the whole cycle of satellite design, development, production, and application. The scope of artificial intelligence in satellite field is very broad, mainly including two aspects. The first is the intelligence of satellites, and the second is the intelligence of satellite design, manufacturing, testing, and satellite data application. We believe that the former has its own characteristics, while the latter has little difference from artificial intelligence in other fields. In the book, we focus on the analysis of satellite artificial intelligence.

The scope of artificial intelligence in the satellite field is very broad, and the content involved at present generally belongs to "+AI," including computational intelligence, multidisciplinary design of space missions, multi-objective optimization of space applications, resource allocation and solutions, intelligent search, and optimization methods in aerospace applications, emerging artificial intelligence technology and group intelligence, intelligent chips, intelligent operating systems, space and in-satellite integrated network, high-performance on-board computer, intelligent algorithm for fault identification, diagnosis and maintenance, multi-agent

Intelligent Satellite Design and Implementation, First Edition. Jianjun Zhang and Jing Li.
© 2024 The Institute of Electrical and Electronics Engineers, Inc. Published 2024 by John Wiley & Sons, Inc.

system and bionic solution for system design and control, machine learning progress of space applications, human–computer interaction intelligent interface, knowledge discovery, data mining and representation of large data sets, etc., that is, traditional industries or currently mature industries actively introduce AI technology to optimize their own business, improve efficiency and user experience, and reduce risks and costs. "+AI" is more about thinking about what technology can do. It is mainly used to transform and optimize the current inherent process, which is normal technology iteration and upgrading [4–6].

In the future, the core of AI to realize intelligence is to learn human's flexible thinking. In the field of satellite, AI will be driven by "AI+" technology as the core and will be launched by satellite systems to redesign products, programs, or work modes with the goal of exploring diversified scene applications. "AI+" tends to think about what technology can do, which may or may not exist at present. Therefore, the logic of "AI+" is more likely to produce "new inventions," which will disrupt the industry.

2.1.1.2 Intelligent Satellite

For the future, intelligent satellite refers to the architecture model simulating the human brain's natural intelligence "information perception, memory thinking, learning adaptation, and action drive," forming a space vehicle with seven self-capabilities of "self-perception, self-memory, self-thinking, self-learning, self-adaptation, self-action, and self-evolution."

At present, there is no consensus on the concept of satellite artificial intelligence. Satellites, including satellites, deep space probes, manned spaceflight, etc., are essentially typical unmanned systems and automatic control machines. Their development is based on the behavioral school theory with cybernetics as the core and develops along the route of automation, autonomy, and high autonomy. Therefore, it can be considered that satellites are essentially unmanned systems with AI attributes and have been a typical representative of AI development for quite a long time. However, in the new wave of AI development, due to the limitations of satellite's own computing resources and data resources, neural networks, deep learning, and other technologies have not been integrated with satellite development, and the development lags behind.

In view of the above discussion, the current main concerns of satellite AI are:

1) Vigorously promote the current deep integration of AI technology represented by big data, machine learning, and high-performance computing with aerospace, and on the basis of automation, enable satellites to have environmental awareness and understanding, autonomous decision-making and planning, self-learning and evolution, autonomous health management, and other functions.

2) In combination with the characteristics of aerospace development, develop the artificial intelligence algorithm theory with aerospace characteristics. Through environmental cognition, decision-making planning, and collaborative action, generate intelligent behaviors with a plan and purpose to adapt to the environment and change the status quo, imitate human intelligence and behavior, and actively perform tasks in complex and changeable unknown environments.

2.1.1.3 Group Intelligence

Group intelligence is the development direction. Intelligent networking satellites can continuously cooperate with many satellites to form an intelligent agent with its own characteristics on the whole, from a lonely intelligent satellite to a distributed intelligent satellite combination [7–10].

Group intelligence originates from the research on the group behavior of social insects represented by bee colonies, ant colonies, etc. (including the information transmitted, the way of information transmission, and how to reach an agreement and take action, that is, decision-making). There is nonsynchronous information exchange between groups. Groups are self-organized, with division of labor and collaborative work. The results of group behavior are neither random nor deterministic; the collective intelligent behavior of a system composed of nonintelligent agents through interaction with each other or with the environment.

In the field of satellites, group intelligence can be expressed as individual aircraft interact with each other in a relatively simple local self-organization manner, and display intelligent characteristics such as distributed, adaptive, and robust in the environment, making the system emerge at the overall level an intelligent level that cannot be achieved by a single aircraft.

2.1.1.4 Understanding of Intelligence and Autonomy

Autonomy and intelligence are two different categories of concepts. Autonomy expresses the way of behavior, and it is called autonomy to complete a certain behavior with its own decision; intelligence refers to the ability to complete the behavior process, that is, whether the methods and strategies used conform to the natural laws or the behavior rules of people (or groups). It is intelligent to find a reasonable "path" to complete a task in an ever-changing environment. Obviously, intelligence is hierarchical.

The relationship between autonomy and intelligence should be: autonomy comes first and intelligence comes second, which should complement each other; autonomy is not necessarily intelligent, but autonomy hopes to be intelligent; intelligence depends on autonomy. The level of intelligence depends on the level of autonomy. Intelligence is the combination of autonomy and knowledge and its use of knowledge. The general process of intelligence generation should be: under the premise of autonomy, comprehensively use the abilities of authority, initiative, love and obsession, perception and other aspects to feel information, extract information, accumulate knowledge, summarize knowledge, summarize the characteristics and refine, improve and perfect the knowledge structure, and integrate the knowledge to achieve the goal of conforming to the laws of nature as much as possible.

Autonomous control refers to the control process of a system to achieve its objectives without human or other system intervention and can adapt to changes in the environment and objects. Classical automatic control methods such as PID, robust, and adaptive control are designed based on analytical models. They are not strong in adaptability to sudden events, unknown environments, and other changes, and their task ability is limited. They belong to a lower level of autonomous control. Intelligent autonomous control refers to autonomous control with human behavior attributes such as perception, learning, reasoning, cognition, execution, and evolution and is the advanced stage of autonomous control.

The intelligence level of satellites is one of the key factors that determine the autonomous capability of satellites. The introduction of artificial intelligence technology improves the intelligence level of space vehicles and enhances the ability of satellites to perform tasks independently. Therefore, intelligence is the attribute, and autonomy is the goal.

2.1.2 Technical Characteristics of Intelligent Satellite

The future space intelligent aircraft has the characteristics of "seven self-capabilities," which can realize its own health protection, independent survival, and evolution and can provide intelligent services for users. Future intelligent satellites have the following typical features at the functional level, implementation level, and apparent level.

2.1.2.1 Functional Level

The satellite itself, the large loop system of satellite heaven and earth, the satellite cluster, and even the space system are essentially information physical systems. Therefore, from the perspective of cybernetics, intelligent satellites should have three basic functions: intelligent perception, intelligent decision-making, and intelligent operation. Here, we call intelligent perception, intelligent decision-making, and intelligent operation the "three elements" of intelligent satellites.

Intelligent perception: According to the multi-source heterogeneous data such as vision, electromagnetic, inertial, astronomy, force touch, etc., on the satellite or other external data sources, obtain the satellite orbit and attitude and other operational status information, its own health status information, and external environment information. Further, through feature extraction, information fusion, and reasoning analysis, it can understand the whole mission scenario and evaluate and predict the situation evolution.

Intelligent decision-making: According to user needs and based on satellite intelligent perception results, it can independently analyze, infer, and predict its own operation rules and dynamic interaction rules with the environment, independently determine task objectives, and carry out task decomposition, planning, and scheduling, and then independently generate task-oriented instruction sequences, the instructions are optimized and updated in real-time according to the execution feedback.

Intelligent operation: According to the task instructions, the interaction feedback and behavioral causality in the interaction between the satellite and the environment can be obtained through learning and other methods, and then the sequence of its own behavior and action can be formed, and then the operation purpose can be achieved through attitude and position control.

The core of realizing the three basic functions of satellite intelligent perception, intelligent decision-making, and intelligent operation is learning. Learning is an important means for satellites to adapt to the uncertainty and unpredictable changes of their internal and external environment and the key to improving satellite performance and the ability to perform complex tasks. It directly affects the intelligence level of satellites. Figure 2.1 shows the relationship between the three elements of intelligent satellite and learning.

2.1.2.2 Implementation Level

Intelligent satellites need to be supported by intelligent architecture, intelligent software, and intelligent hardware with deep integration of information and physics.

Intelligent system architecture: It covers the information network system applicable to massive satellite data storage, high-speed transmission and interaction,

Figure 2.1 Schematic diagram of the relationship between the three elements of intelligent satellite and learning.

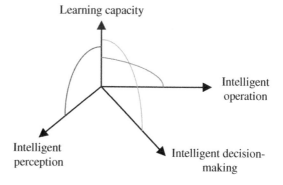

and deep fusion processing, as well as the logic system convenient for realizing intelligent perception, intelligent decision-making, and intelligent operation functions. It can have an open structure, which is conducive to the deep interaction and integration of information systems and physical systems.

Intelligent software: Computer software that endows the satellite with intelligent behavior, which can realize functions such as knowledge processing, problem solving, and on-site induction (environment adaptation). It covers intelligent operating systems, artificial intelligence programming language systems, intelligent software engineering support environments, intelligent human–computer interface software, intelligent expert system, intelligent application software, and others.

Intelligent hardware: The hardware facilities endowed with satellite intelligent behavior can realize high-performance computing, high-speed data exchange, intelligent perception, intelligent operation and other functions, covering intelligent computers, intelligent buses, intelligent perception devices, intelligent control devices and other types. Through the loading of Internet services, the "cloud + end" application architecture can be formed.

2.1.2.3 Apparent Level

Intelligence has diversity: The "humanoid skills" possessed by intelligent satellites have diversity; through intelligent empowerment, satellites can have perception intelligence, decision intelligence, and operation intelligence, and based on this, multiple types of behavioral intelligence using knowledge to solve problems can be derived.

Intelligence has multiple layers: Aerospace systems have multiple layers in terms of structure, information interaction, etc. From load, platform to the satellite system, and then to the large loop system of heaven and earth, each layer can have different levels of intelligence, and the different levels are closely connected and relatively independent. The intelligence of individual or group of satellites can be realized by maximizing the intelligence of each level.

2.2 Technical Characteristics of Intelligent Satellite System

Under the background of the development of the new generation of artificial intelligence in full swing, the analysis and study of the classification and technical architecture of satellite intelligence can clarify the development stage of the current satellite intelligence technology, sort out the technical system of the future development of intelligent satellites, and provide reference and support for the development planning of intelligent satellites.

2.2.1 Intelligent Classification of Satellite System

The diversity and multi-layer of intelligent satellites determine that their intelligence level is diverse. The following is a comprehensive analysis of the existing satellite intelligent autonomous control classification and satellite intelligent reasoning classification.

The intelligent classification of satellite systems includes intelligent satellites, intelligent payload, intelligent service, and intelligent manufacturing. Here, only the intelligent classification of intelligent satellites is given. The classification of intelligent load, intelligent service, and intelligent manufacturing needs to be improved [11–13].

2.2.1.1 Classification of Satellite Intelligent Autonomous Control

1) Three grades of ANTSAKLIS in 1994

In 1994, ANTSAKLIS and others divided the general control system into the following three levels according to whether they can complete control, perception and decision-making independently:

Level 1: Be able to obtain environmental information and make decisions and controlled movements.

Level 2: Be able to identify targets and events, express knowledge, reason, and predict the future.

Level 3: Be able to perceive and understand the environment, make optimal decisions, and carry out controlled movement under a wide range of changes in the environment, so as to survive and grow in a dangerous hostile environment.

2) NASA's four levels in 2002

In 2002, NASA classified the intelligence level of the aircraft control system from three aspects: the aircraft's ability to adapt to the environment, the ability to optimize performance indicators, and the ability to respond to emergencies, which are divided into the following four levels:

Level 0: Under nominal conditions, the aircraft can track the desired trajectory (robustness)

Level 1: Under nonnominal conditions, the aircraft can track the desired trajectory, and the controller can actively adapt to environmental changes (robustness + adaptability)

Level 2: Optimize a performance index on the premise of tracking the desired trajectory. (robust + adaptability + optimization)

Level 3: Be able to carry out independent planning, respond to emergencies, and have the ability to fault diagnosis and recovery. And these abilities have self-adjustment to adapt to environmental changes.

The above classification mainly classifies the robustness and adaptability of the aircraft attitude and trajectory control system but is still limited to the traditional basic design framework of navigation, guidance, and control.

2.2.1.2 Classification of Satellite Intelligent Reasoning

In 2006, the Space Operations and Support Technology Committee (SOSTC) of the American Academy of Aeronautics and Astronautics (AIAA) made a more comprehensive division of the level of integrated intelligent autonomy of space vehicles. They summarized the relevant information of 88 different autonomous intelligent space robots/satellite systems from 12 different organizations around the world. In the book, autonomy/intelligence is divided into six levels according to the level of intelligent reasoning, as shown in Table 2.1. With the gradual improvement of these six levels, the autonomy and intelligence of satellites will also be improved [14–17].

The above classification divides the level of intelligent autonomy according to the degree of human participation in the system. The "strong intelligence" development goal of intelligent reasoning and independent thinking in orbit is proposed.

2.2.2 Intelligent Satellite Technology Architecture

Intelligent satellite system refers to the satellite system after intelligent empowerment. The technical category of artificial intelligence in the satellite field includes intelligent satellites, intelligent payload, intelligent services, and intelligent

Table 2.1 Classification of intelligent reasoning/intelligent control by AIAA.

Grade	Ability	Describe
1	Manual operation	—
2	Automatic notification	Automatic interpretation functions such as threshold detection are introduced on the basis of program control
3	Manned ground intelligent reasoning	Introduce expert knowledge to give implementation suggestions for external events
4	Unmanned ground intelligent reasoning	The ground system automatically calculates the command sequence and executes it
5	On-track intelligent reasoning	On-track calculation and execution of instruction sequences and the ability to respond to external events
6	Self-thinking satellite	Be able to independently analyze task targets, calculate action instructions, and execute

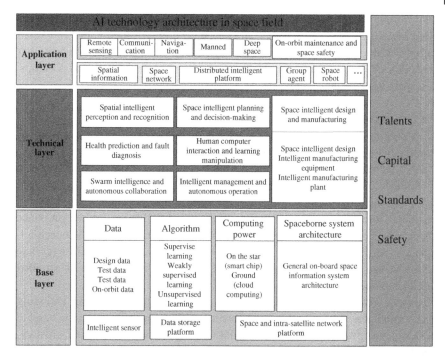

Figure 2.2 Artificial intelligence ecology in satellite field.

manufacturing. Here, we will only discuss the technical scope of intelligent satellites in-depth, and the technical scope of intelligent loads, intelligent services, and intelligent manufacturing needs to be improved [18–20].

The artificial intelligence technology system in the satellite field includes three parts: basic layer, technical layer, and application layer, as shown in Figure 2.2.

2.2.2.1 Basic Layer

The basic layer directly faces the technical basis of AI application in space, which is mainly composed of data, algorithm (model and software), computing power (AI chip), and on-board operating system. The aerospace big data includes the data accumulation formed in the process of satellite manufacturing, testing, and operation, which lays the foundation for the realization of artificial intelligence in the aerospace field; the algorithm mainly includes intelligent algorithm model based on small samples; computational power mainly refers to intelligent chips (AI chips: GPU, FPGA, NPU, etc.) suitable for aerospace applications, which are responsible for computing; the on-board operating system refers to the general on-board space information system, which has high-performance in-orbit information processing capability.

According to the characteristics of aerospace, the development focus of the basic layer includes the theory of efficient learning with high value and small samples under resource-constrained conditions, the theory of interpretability and reliability of deep learning intelligent algorithms, high-performance AI chips, aerospace big data, and intelligent on-board information systems.

2.2.2.2 Technical Level

According to the characteristics of satellites, artificial intelligence technology in satellite field can be divided into two categories.

First, technologies related to the normal operation and survival of satellites, including spatial intelligent perception and recognition, spatial intelligent planning and decision-making, health prediction and fault diagnosis, human–computer interaction and learning control, group intelligence and autonomous cooperation, intelligent management, autonomous operation, etc. These technologies can significantly improve the viability of satellites and improve the serviceability to users. Its breakthrough will contribute to the formation of artificial intelligence theory with aerospace characteristics.

Second, technologies related to satellite design and manufacturing, including space intelligent design and manufacturing, can optimize the work of satellite design, development and verification, and create distributed, integrated, efficient, and high-quality aerospace characteristic products.

2.2.2.3 Application Layer

The application layer mainly uses artificial intelligence technology to realize certain space missions, such as intelligent application satellites (communication, guidance, and remote control), intelligent manned space, intelligent space exploration, intelligent space security and on-orbit maintenance, and includes multi-system integration applications, such as space-integrated space information, space network, distributed intelligence, and group intelligence. The application layer is the direct embodiment of AI technology and the fundamental purpose of developing AI.

2.3 Opportunities and Challenges for Satellite to Develop AI

2.3.1 Development Opportunities of Smart Satellite

AI is a strategic technology leading the future. As the core driving force of the new round of industrial transformation, AI will further release the huge energy accumulated by previous scientific and technological revolutions and industrial transformations. The development of space artificial intelligence technology will surely

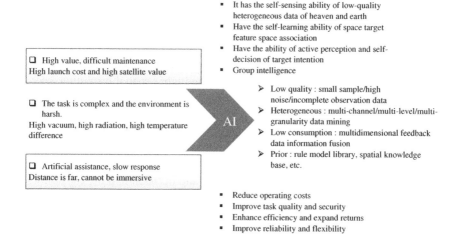

Figure 2.3 Opportunities for the development of artificial intelligence in satellite field.

lead to a chain breakthrough and accelerate the development of satellites to a new generation of intelligence [21–23].

In the future, with the development of cutting-edge new technologies such as brain science, neural computing, quantum computing, biological intelligence, etc., it is bound to inject new vitality into the intelligent development of satellites. It may bring disruptive breakthroughs and development in the design concept, system architecture, operation mechanism, and service mode of satellites, as shown in Figure 2.3.

2.3.1.1 AI Leads the Disruptive Innovation of Future Satellite Platforms Is an Inevitable Trend

During the development, verification, testing, flight control, and delivery of satellites, the problems such as unattended in the target space environment, high-reliability requirements, high cost of testing and maintenance, and many failure factors have been puzzling to scientific researchers. Artificial intelligence supporting satellite system technology is a powerful means to solve these problems. It provides a rich technical reserve for improving satellite intelligence autonomous ability. It is one of the primary development directions of intelligent satellite platform design in the next decade. Space technology will realize the leap from automation to autonomy to intelligence.

2.3.1.2 Artificial Intelligence Technology Will Effectively Improve the Efficiency of Satellite Systems

Artificial intelligence technology is the key technology to endow space unmanned systems with intelligent properties. The application and integration of artificial intelligence technology in various existing satellite systems will effectively improve the level of intelligence and autonomy of satellites and significantly improve the efficiency of existing satellite systems.

1) Effectively improve the intelligent perception and recognition ability of the satellite system. In the future, while pursuing higher spatial resolution, spectral resolution and time resolution, optical remote sensing, navigation, and communication satellites need to develop their data intelligent acquisition capabilities, image data intelligent processing and information extraction capabilities, as well as the perception and recognition capabilities for various tasks to improve the application efficiency of remote sensing satellites.

2) Effectively improve the intelligent planning and decision-making capabilities of satellite platforms. Future on-orbit module replacement, fuel filling, on-orbit assembly, unknown environment detection, and other complex on-orbit operations and services, manned and deep space missions require unmanned space systems to have "human-like" flexible, intelligent behaviors such as perception, understanding, decision-making, and execution.

3) Improve the ability of intelligent group games on satellite platforms in all aspects. The development of air and space combat equipment represented by intelligent space countermeasure weapon systems, intelligent space-based trans-domain strike weapon system, and intelligent space-to-earth unmanned combat weapon systems is an important development direction in the field of air and space attack and defense field in the future. The development of intelligent space attack and defense systems will be of great significance for improving space attack and defense capability and mastering the initiative of future space battlefields.

4) Improve the intelligent space detection capability of satellite platforms by multiple means. Facing the complex exploration tasks on the surface of the moon/ Mars and other extraterrestrial objects in the future, the requirements for the dexterity, intelligence, and autonomy of the detector/robot are increasingly high. It needs to have intelligent autonomous operation capabilities such as autonomous path planning, intelligent obstacle avoidance, autonomous observation, intelligent identification of interested objects, autonomous sampling, and so on in an unstructured and unfamiliar environment. Therefore, it is necessary to combine artificial intelligence technology to develop the intelligent detection technology of the surface detector/robot of extraterrestrial objects.

2.3.2 Significance of Intelligent Satellite

AI will be a subversive force in the field of national security, and its impact can be compared with nuclear, aerospace, information, and biotechnology. The space power led by NASA of the United States has taken space as an important stage for AI to play its role [24–26].

2.3.2.1 The Application and Development of Artificial Intelligence Technology in the Satellite Field Closely Meet the Needs of the Times

1) The development of artificial intelligence in the aerospace field is an urgent requirement to innovate the future war form and win the future war. With the rapid development of the new military revolution, war is changing to the intelligent form, and the intelligence of weapons and equipment has become the hot spot of the development of the major military countries. Artificial intelligence has been used as the new power to accelerate the development of weapons and equipment to prevent the risk of further widening the gap between the performance of weapons and equipment and that of developed countries.

2) The development of artificial intelligence in the aerospace field is the only way for the development of aerospace equipment. Faced with the characteristics of noncooperation, borderless, and incomplete information of aerospace equipment, as well as the requirements of exploration, uncertainty, and strong confrontation in the future battlefield and space environment, the demand for intelligent technology of aerospace equipment is more urgent than that of other industries.

In short, the successful application of AI in various fields has laid a good foundation for the design of satellite systems and the development of satellite intelligence in the future. Without intelligent autonomy in the field of deep space exploration, it will be impossible to move forward in the future; manned satellites begin to develop and configure intelligent robot assistants in sealed cabins to assist astronauts' on-orbit activities; the in-orbit service satellite will develop a noncooperative satellite capture system with computer vision and cognitive reasoning as the core; intelligent autonomous operation or supervised autonomous operation will become an important development direction of the future space system; the development of intelligent military aerospace integrates many cutting-edge intelligent technologies, and so on.

2.3.2.2 Artificial Intelligence Technology Will Open Up a New Working Pattern of Integrating Space and Earth in Satellite Field

In the process of coordinated development between space and earth, on the one hand, while researching satellite intelligence, study the ground system intelligence, including ground intelligent command and control decision-making, ground intelligent human–computer interaction, etc., and promote the coordinated development of integration of space and earth of intelligent space attack and defense system in the same and cross-domain.

On the other hand, satellite and ground control work can be "intelligent and autonomous" and achieve "integrated design" to improve the operability of flight and ground interaction work, that is, "integrated design," which is an issue that can be raised to a strategic height and can reduce the construction costs and risks of each other.

The satellite is an unmanned autonomous system that is subject to space characteristics. The current task cannot achieve intelligent autonomy, and the working mode is "autonomous + remote control." In the face of the future situation and task requirements, the star cluster cooperates with each other, and the information data can be accurately recognized and planned for the first time, and the intelligent control behavior can be accurately executed, which becomes the top priority. We should develop intelligent means such as learning and evolution in the satellite field and face the new working mode "constellation + co-domain/cross-domain + collaboration + space and earth."

2.3.2.3 The Development of Artificial Intelligence Technology Will Comprehensively Promote the Upgrading of the Ecological Industrial Structure in the Satellite Field

Artificial intelligence technology will become the catalyst and core support for comprehensively promoting the upgrading and upgrading of the ecological industrial structure in the satellite field in the new era. It will drive technological development and innovation from the satellite system preresearch, design, development and production, integration testing, test verification, in-orbit operation, and application and other stages.

It can strengthen the linkage up and down the industrial chain, optimize the technology and management system, realize the coordinated development of data, algorithms, computing power, technology, application, ecology, etc., strengthen the key technology tackling, and optimize the industrial development environment. Actively promote production, learning, research, and application to strengthen technical research and encourage the exploration of new technological paths. Promote the gathering of talents, funds, resources and other factors, and effectively promote the landing of artificial intelligence in the satellite field.

2.3.3 Major Challenges of Intelligent Satellites

Compared with most ground projects, the current AI technology represented by deep learning faces many challenges when applied on satellites.

1) The combination of space technology and artificial intelligence aims to reshape the space system and create a new application scenario for satellites.

   The combination of space technology and artificial intelligence is aimed at reshaping the space system, creating new application scenarios for satellites, interpreting and recognizing the integration of "communication, guidance, and remote" fields, the tacit cooperation of space groups, the new mode of space attack and defense and security enabled by intelligence, and the efficient exploration of the unknown environment of the integration of space and earth. In the next 10–20 years, AI technology will be widely and rapidly applied in the field of satellites, which will have a profound impact on the development of the new generation of flexible space system construction, autonomous sensing, control and intelligent cooperation technology, real-time detection and advanced space command, and control technology.

2) Reconstruct the function and performance of satellite platforms based on artificial intelligence, and face future elastic changes and demands.

The application and integration of artificial intelligence technology in various existing satellite systems will effectively improve the level of intelligence and autonomy of satellites and significantly improve the efficiency of existing satellite systems. Through the development of artificial intelligence in the field of satellites, we will coordinate the system engineering, overall, professional and satellite development and operation stages, give full play to our comprehensive advantages, form internal and external forces, comprehensively improve the function and performance of the satellite platform, and face the future elastic changes and demands.

The development of satellite platforms in the future has the following trends: the basic ability of the platform to perform space tasks tends to mature, and the intelligent and autonomous multi-party response ability needs to be developed, which requires the satellite to have the autonomous survival ability in orbit; higher requirements for timeliness, safety, and stability of satellite platform system control; the application capabilities for complex multi-load detection, game against space attack and defense, and on-orbit control of non-cooperative targets are oriented to all-round improvement; the tasks are increasingly complex and changeable, and the platform needs to have the basic capabilities of learning generalization, aggregation and distribution, cluster collaboration, efficiency and reliability, etc.

3) Satellite platform energy, computing power, data, and other resources are limited, and it is necessary to develop algorithms and computing power that adapt to the space environment.

The hardware layer requires cutting-edge hardware devices. The combination of aerospace and artificial intelligence, in addition to the design of intelligent algorithms at the software level, the most significant difference is the difference between artificial intelligence support systems. Unlike ground support systems, on-board systems have many limitations on processing performance, transmission speed, storage capacity, and energy supply.

The software layer needs efficient new algorithms. At present, the use of deep reinforcement learning technology for learning control of space tasks has shown great potential, but there are still many challenges in terms of expansion and stability.

4) The development of new software and hardware design environment and system architecture requires the introduction of artificial intelligence and related technologies.

   Directly go deep into "the first use of cutting-edge technological achievements in aerospace," form the theory of intelligent aerospace engineering, reasonably plan the technology application mode from the top architecture, and carry out the research on the design of on-board intelligent chips, system design based on artificial intelligence, on-orbit fault detection, and maintenance based on artificial intelligence, satellite intelligent control based on artificial intelligence, satellite-ground integration based on artificial intelligence and satellite intelligent platform based on deep reinforcement learning, etc. We will vigorously develop the cutting-edge technology of space intelligence software, hardware, and system from point to point and support the future of intelligent aerospace.

5) The aerospace industry has high requirements for safety and reliability, and the current AI technology has interpretability problems.

The AI algorithm represented by deep learning still has many problems that are criticized by people, such as its convergence, interpretability, and reliability. Many researchers are exploring more efficient machine learning methods that can replace deep learning. Hinton, the father of deep learning, also made a subversive statement: "Deep learning should start anew and completely abandon reverse propagation." The challenges of applying AI technologies such as deep reinforcement learning to the aerospace field are as follows:

1) *Stability, robustness, and interpretability*: Deep reinforcement learning (DRL) algorithms may be relatively unstable, and their performance may vary greatly among different configurations. In order to solve this problem, a deeper understanding of the learned network representation and strategy may help to detect the scenario of confrontation to prevent the satellite system from being threatened by security.

2) *Lifelong learning*: At different times and in different scenes, the environmental perception of space probe navigation or space station manipulator operation will change, which may hinder the implementation of the learned control strategy. Therefore, the ability to continue learning to adapt to environmental changes and the ability to maintain solutions to the environment that has been experienced is of key value.

3) *Generalization between tasks*: At present, most algorithms are designed for a specific task, which is not ideal because the intelligent satellite system is expected to complete a group of tasks and conduct the minimum time of total training for all considered tasks.

With the continuous deepening of research on artificial intelligence theories and methods represented by DRL, human beings will achieve the goal of "solving intelligence and solving everything with intelligence" soon.

References

1 Vladimirova, T. and Wu, X.F. (2006). On-board partial run-time reconfiguration for pico-satellite constellations. In: *Adaptive Hardware and Systems, 2006 (AHS 2006), First NASA/ESA Conference on, 2006*, 262–269. IEEE.

2 Marzwell, N.I., Waterman, R.D., KrishnaKumar, K., and Waterman, S.J. (2005). How to extend the capabilities of space systems for long duration space exploration systems. In: *AIP Conference Proceedings*, vol. 746, 1153–1162. American Institute of Physics.

3 Crawford, B.S. (1968). Operation and design of multi-jet spacecraft control systems. PhD. thesis. Cambridge: Massachusetts Institute of Technology.

4 Servidia, P.A. (2010). Control allocation for gimballed/fixed thrusters. *Acta Astronautica* 66 (3-4): 587–594.

5 Bayard, D.S. (2001). An optimization result with application to optimal spacecraft reaction wheel orientation design. In: *Proceedings of the American Control Conference*, Arlington, VA, 1473–1478. IEEE.

6 Fleming, A.W. and Ramos, A. (2012). Precision three-axis attitude control via skewed reaction wheel momentum management. In: *Paper no. 79-1719, AIAA Guidance and Control* (6–8 August, Boulder, CO), 177–190. New York: AIAA.

7 Bordignon, K.A. (1996). Constrained control allocation for systems with redundant control effectors [D]. PhD thesis. Blacksburg: Virginia Polytechnic Institute and State University. pp. 17–19, 21, 67–83, 85–102.

8 Lappas, V., Richie, D., Hall, C. et al. (2009). Survey of technology developments in flywheel attitude control and energy storage systems. *Journal of Guidance, Control, and Dynamics* 32 (2): 354–365.

9 Richie, D.J., Lappas, V.J., and Prassinos, G. (2009). A practical small satellite variable-speed control moment gyroscope for combined energy storage and attitude control. *Acta Astronautica* 65: 1745–1764.

10 Burken, J.J., Lu, P., and Wu, Z.L. (1999). Reconfigurable flight control designs with application to the X-33 vehicle. In: *AIAA Paper No. 99-4134. AIAA Guidance, Navigation, and Control Conference*, Portland, OR, 951–965. AIAA.

11 Luo, Y., Serrani, A., Yurkovich, S. et al. (2007). Model-predictive dynamic control allocation scheme for reentry vehicles. *Journal of Guidance, Control and Dynamics* 30 (1): 100–113.

12 DARPA (2016). RSGS program solicitation DARPA-PS-16-01. USA: DARPA.

13 An, C.C., Kiminura, A., Barrios, L. et al. (2014). Dynamic power sharing for self-reconfiguration modular robots. *Lecture Notes in Computer Science* 1: 3–14.

14 Taylor, G.J., Lawrence, S., Lentz, R. et al. (2005). Strategies for using SperBpts on the Lunar surface. [C/OL]. http://www.lpi.usra.edu/meetings/leag2005/ (accessed 08 September 2015).

15 Barnhart, D., Hill, L., Foeler, E. et al. (2013). A market for satllite cellularization: a first look at the implementation and potential impact of Satlets. In: *Proceedings of AIAA Space 2013 Conference and Exposition*, 2579–2589. Washington DC: AIAA.

16 Aerospace Corporation (2018). Hive satellites redefine disaggregation. [EB/OL]. https://aerospace.org/article/hive-satellites-redefine-disaggregation (accessed 08 February 2018).

17 Hill, L., Foeler, E., Jaeger, T. et al. (2013). DARPA Phoenix Satlets: progress towards satellite cellularization. In: *Proceedings of AIAA Space 2013 Conference and Exposition*, 2579–2589. Washington DC: AIAA.

18 German Aerospace Center (2017). i BOSS–Intelligent building blocks for on-orbit satellite servicing and assembly. [EB/OL]. https://www.iboss.space/wp-content/uploads/2017/09/IAC-17D123x40674.pdf (accessed 25–29 September 2017).

19 Ye, Z., Wang, S.H., and Zhao, T.D. (2017). IMA dynamic reconfiguration modeling and resource criticality analysis based on Petri net. In: *2017 Second International Conference on Reliability Systems Engineering (ICRSE)*, 1–6. IEEE. ISBN: 978-1-5386-0918-7.

20 Campbell, M.E. (2005). Collision monitoring within satellite clusters. *IEEE Transactions on Control Systems Technology* 13 (1): 42–55.

21 Kruk, J.W., Class, B.F., Rovner, D. et al. (2003). FUSE in-orbit attitude control with two reaction wheels and no gyroscopes. In: *Future EUV/UV and Visible Space Astrophysics Missions and Instrumentation*, vol. 4854, 274–285. SPIE.

22 Sahnow, D.J., Kruk, J.W., Thomas, B. et al. (2006). Operations with the new FUSE Observatory: three-axis control with one reaction wheel. In: *SPIE Astronomical Telescopes + Instrumentation*, 6266. Proceedings of SPIE - The International Society for Optical Engineering.

23 Pelene L. (2006). SPACEBUSTM, a vehicle for board missions. AIAA-2006-5303.

24 Ruiter, A.D. (2013). A fault-tolerant magnetic spin stabilizing controller for the JC2Sat-FF mission. *Acta Astronautica* 68 (1-2): 160–171.

25 Morgan, P.S. (2012). Resolving the difficulties encountered by JPL interplanetary robotic spacecraft in flight. In: *Advances in Spacecraft Systems and Orbit Determination* (ed. R. Ghadawala), 236–264. IntechOpen.

26 Brown, T. (2012). In-flight position calibration of the Cassini Articulated Reaction Wheel Assembly. In: *AIAA Guidance, Navigation, & Control Conference*, Minneapolis, Minnesota (13–16 August). AIAA.

3

Development Status of AI Technology in Satellites

As the core driving force of the new round of industrial transformation, AI has a strong enabling role for traditional industries while accelerating the birth of new technologies and new products. It can trigger major changes in the economic structure, form a development pattern of AI replacing low-end sectors and complementing high-end sectors, realize the overall leap of social productivity, and move from +AI to AI+ close to human flexible thinking. Governments and ministries of defense around the world have issued policies and plans around AI to deploy core AI technologies, top talents, standards, and specifications and accelerate the development of AI technology and industry [1–3].

To seize the commanding heights of the future space, advanced space countries have increased their investment in intelligent technology in their space plans and launched relevant AI development plans. There is fierce competition between space intelligent perception and recognition (including navigation, etc.), space intelligent planning and decision-making (including control, etc.), health prediction and fault diagnosis (including operation management, etc.), human–computer interaction and learning control (including heaven and earth, etc.), group intelligence and autonomous cooperation intelligent management and autonomous operation (including data management, etc.).

3.1 Policy and Planning

The United States has always been at the forefront of AI research in the world. The United States government has played a key role in AI research, and the Ministry of Defense, the Ministry of Commerce, and NASA NSAS have also actively participated in AI-related research and development. The European

Intelligent Satellite Design and Implementation, First Edition. Jianjun Zhang and Jing Li.
© 2024 The Institute of Electrical and Electronics Engineers, Inc. Published 2024 by John Wiley & Sons, Inc.

Union, Germany, France, the United Kingdom, Japan, Canada, Russia, and other governments and defense departments have also introduced AI development strategies. See Table 3.1 for the main policies and regulations issued by advanced aerospace countries [4, 5].

It can be seen from Table 3.1 that the focus of the AI policy issued by the United States is to respond to the general trend of AI's vigorous development and focus on the long-term impact and change on national security and social stability. At the same time, the United States, as a leading power in science and technology, maintains its initiative and foresight in the development of AI, for important AI fields (chips, operating systems, military, energy, etc.); it strives to maintain its leading position in the world. The European Union will take a three-pronged approach to promote the development of artificial intelligence in Europe. The European Space Agency stated that information technologies such as cloud computing, big data, artificial intelligence and machine learning have become emerging technologies that change the rules of the game in the field of Earth observation. It is necessary to study the application prospects of block-chain technology in the field of Earth observation in advance to achieve rapid and widespread application. The German Federal Government believes that AI has entered a new stage of maturity, and that "AI made in Germany" should become a globally recognized product logo. France plans for the development of artificial intelligence in the future, aiming to make France the leader of artificial intelligence in Europe. The British government has analyzed the current application, market and policy support of AI, and put forward important action suggestions to promote the development of AI industry in the UK from four aspects: data acquisition, talent training, research transformation, and industry development. Japan has listed unmanned systems on the ground, in the air and at sea as the key areas of military technology development, and has taken unmanned technology and intelligent technology as the key direction of military technology development. Canada, Russia, South Korea, Singapore, the United Arab Emirates and other countries have also issued their respective AI policy plans [6, 7].

3.2 Technology and Application

AI includes all technologies that enable computers to imitate intelligence. Due to the limitations of computing, storage, transmission, and other hardware levels, the most successful AI implementation based on DL *(deep learning) is* rarely used in today's aerospace industry. In addition, the (statistical) model developed by the training neural network is not human-readable and *cannot be reproduced.* AI, especially ML *(machine learning)*, has a long way to go before it is widely

Table 3.1 Relevant policies and regulations issued by advanced aerospace countries.

Serial number	Country	Policies and regulations	Sending time	Issuing unit	Primary coverage
1.	United States	National Strategic Plan for Artificial Intelligence Research and Development	2016.10	National Science and Technology Committee (NSTC)	*Set the development of AI as a national strategy and determine seven long-term strategies:* long-term investment in AI research and development; develop effective methods of human–computer cooperation; understand and respond to the ethical, legal, and social impacts of AI; ensure the security of AI system; develop AI shared data set and test environment platform; establish standards and benchmarks to evaluate AI technology; better grasp the national AI R&D talent demand
2.		Preparing for the Future of AI, AI, Automation, and Economics Report	2016.12	Executive Office of the President, NSTC	Dealing with the AI-driven automation economy is a major policy challenge for the government. The development of AI releases the creative *potential* of enterprises and workers and needs to ensure the leadership of the United States in the *creation and use of AI*
3.		A new version of the National Defense Strategy and a contract worth $885 million	2018.01	Ministry of Defense, Defense Contractor Booz Allen Hamilton	Publish its abstract "2018 National Defense Strategy Summary – Strengthening the *Competitive Advantage* of the US Army." *A secret AI project* will be developed in the next 5yr. The specific content is not disclosed
4.		NASA's Frontier Development Laboratory (FDL) Statement	2018.10	NASA	NASA said that the *five ways of artificial intelligence* (making *intelligent predictions of extraterrestrial life, detecting exoplanets, helping* to protect the Earth from asteroids, helping to recover meteorites, and drawing lunar craters that may contain water) play a huge role in human exploration of the universe

(*Continued*)

Table 3.1 (Continued)

Serial number	Country	Policies and regulations	Sending time	Issuing unit	Primary coverage
5.		Sign an executive order to launch the American AI Initiative	2019.02	President Trump	It aims to mobilize more federal funds and resources for *AI R&D* from the national strategic level to meet the challenges from "strategic competitors and foreign competitors" and ensure leadership
6.		Department of Defense AI Strategy Summary	2019.02	Ministry of Defense	Using AI to promote our security and prosperity, the strategy sets the goal of supporting military personnel and protecting the country, led by the *Joint Artificial Intelligence Center (JAIC). The military will invest up to $2 billion in the next 5yr*
7.	European Union	EU AI	2018.04	European Commission	The report proposes that the EU will take a *three-pronged approach to promote the development of artificial intelligence in Europe*
8.		AI Coordination Plan	2018.12		Through seven specific actions, Europe will become a *global leader in the development and application of artificial intelligence,* and ensure that the development of artificial intelligence always follows the principle of "human-centered" and always conforms to the ethical code
9.	Europe	Statement of the European Space Agency (ESA) Office of Advanced Concepts and Research	2018.10	European Space Agency (ESA)	The statement said that AI had changed the rules of the game to make scientific research and exploration more efficient. AI not only doubles this efficiency but AI *makes space exploration 10 times more efficient.*

No.	Country	Title	Date	Issuing Body	Content
10.		Blockchain and Earth Observation White Paper	2019.04		Information technologies *such as cloud computing, big data, artificial intelligence, and machine learning* have become emerging technologies *that* change the rules of the game in the *field of Earth observation. It* is necessary to study and judge the application prospects of *blockchain technology* in the field of Earth observation in advance to achieve rapid and widespread application
11.	Germany	Key Points of Innovation Policy	2017	Ministry of Economy	To occupy the world's leading position in science and technology in the future, we will strengthen the cooperation between industry, university, and research, develop Industry 4.0, and focus on the development of microelectronics, *artificial intelligence*, biotechnology, and quantum technology
12.		German Federal Government's AI Strategy Report	2018.11	Federal Government	*AI has entered a new stage of maturity. We should make "AI Made in Germany" a globally recognized product logo.* The report covers the development and wide application of AI and the possible changes in politics, economy, culture, security, law, ethics, international cooperation, and other aspects it brings and puts forward action measures
13.		Research and development funding of artificial intelligence and talent competition	2019		The German government plans to invest 3 billion euros by 2025 to ensure competitiveness based on good research and innovation policies. Germany plans to add at least 100 professor positions in artificial intelligence and establish an ambitious young talent plan
14.	France	Comprehensive Report on French AI Strategy	2017.03	Ministry of National Education, Higher Education and Scientific Research and Ministry of Economy and Finance	With the goal of ensuring France's *leading position in the field of artificial intelligence*, education and talent cultivation, commercial application, and industrialization, enabling artificial intelligence to develop "vertically" in a suitable economic and ecological environment, and promoting the public's understanding of *artificial intelligence, we plan the future development of artificial intelligence in France*

(Continued)

Table 3.1 (Continued)

Serial number	Country	Policies and regulations	Sending time	Issuing unit	Primary coverage
15.		Artificial Intelligence and Innovation Roadmap	2018.03	Ministry of Defense	Including human-machine cooperation research
16.	UK	Developing AI in the UK	2017.10	British Government	In the report, the current application, market and policy support of AI are analyzed, and important action suggestions to promote the development of AI industry in the UK are put forward from four aspects: data acquisition, talent training, research transformation and industry development. The report was included in the British Government's 2017 White Paper on Government Industry Strategy Guidance, and became an important guide for the development of AI in the United Kingdom
17.		New Deal for AI Industry	2018.04		The report covers the promotion of government and company research and development, STEM education investment, improvement of digital infrastructure, increase of AI talents and leadership of global digital moral exchange, and other aspects, aiming to promote the UK to become a *global AI leader*
18.	Japan	Defense Technology Strategy and Medium and Long-term Technology Plan	2016.08	Japanese Government	The unmanned system on the ground, in the air and at sea is included in the key development of military technology, and the *unmanned technology and intelligent technology are regarded as the key direction of military technology development*. This is the first time Japan has released a top-level strategic document for military technology development after World War II

19.	Artificial Intelligence Industrialization Roadmap	2017		*It is planned to promote the use of AI technology in three stages* to significantly improve the efficiency of manufacturing, logistics, medical, and nursing industries. The first stage (around 2020) is to establish the technology of no artificial factory and no farm, Popularize the use of artificial intelligence for drug development support, Anticipate production equipment failures through artificial intelligence. In the second stage (2020–2030), complete unmanned transportation and distribution of personnel and goods will be realized; multi-energy and coordination of robots; realize drug development for individuals; use of artificial intelligence to control home and home appliances. In the third stage (after 2030), the nursing robot becomes a member of the family, popularize mobile automation and unmanned "to reduce human death accidents to zero"; analyze potential consciousness through AI and visualize "what you want."
20.	Pan-Canada AI Strategy	2017.03	Federal Government of Canada	Cultivate *AI researchers on a large scale; establish three AI research centers* in Montreal, Toronto, and Edmonton; study the economic, ethical, policy, and legal issues brought by the progress of AI; support the establishment of a national research network for artificial intelligence in Canada
21.	The Concept of Developing Military Science Complex by 2025 and the National Weapons and Equipment Plan from 2018 to 2025	2017	Ministry of Defense	To guide the research, development and use of unmanned combat equipment of the Russian army

(Continued)

Table 3.1 (Continued)

Serial number	Country	Policies and regulations	Sending time	Issuing unit	Primary coverage
22.		National Strategy for Artificial Intelligence	2018.06	Russian government	Encourage private enterprises and organizations to strengthen cooperation with multiple government agencies in AI.
23.	Korea	Exotrain Plan	2013.05	Korean government	Develop a natural language dialogue system for human–computer communication in the professional field, with a total budget of US $90 million
24.	Singapore	《AI.SG》	2017.05	National Research Foundation (NRF)	It is planned to invest US $107 million in the next five years to support AI.SG plan, to *use AI* to solve major challenges faced by society and industry, invest in *AI "deep" capabilities* to catch the next wave of technological innovation, and expand the application of AI and machine learning in the industrial field
25.	The United Arab Emirates	UAE AI Strategy	2017.10	Government of the United Arab Emirates	*AI* is widely used in the government and the private sector to improve efficiency, stimulate economic vitality, and build the UAE into a technology and legislative center of AI

used in space applications, but we have begun to apply it to new technologies. *One of the most successful areas of AI application is satellite operation*, especially in supporting the operation of large satellite constellations, including relative positioning, communication, end-of-life management, etc. ML systems are also commonly used in space applications to approximate the complex representation of the real world. For example, ML plays an important role in analyzing many Earth observation data or telemetry data from satellites. In addition, it is found that ML systems that analyze large amounts of data from each space mission are becoming more and more common. Some Mars probe data are being transmitted using artificial intelligence, and these mobile systems have even been taught how to navigate by themselves. In the past decades, its development has gone a long way, but ML needs complex models and structures to be widely used. AI currently lacks the reliability and adaptability required by new software, and these qualities need to be improved before taking over the aerospace industry [7–9].

Although complex technologies such as deep learning (DL) are difficult to apply directly in space systems, some studies have begun to study the application of AI in space applications and satellite operations.

3.2.1 NASA

In terms of spatial intelligent perception and recognition: In the US Hyperspectral Remote Sensing Technology Program (HRST), real-time adaptive spectral recognition system (ORASIS) is used for real-time processing and compression of onboard data, which can perform automatic data analysis, feature extraction, and data compression. The ASE software system on the Earth Observer 1 (EO-1) satellite can enable the satellite to carry out independent scientific exploration and respond to scientific events on the earth, and so on [10, 11].

In terms of space intelligent planning and decision-making: The AEGIS system on Curiosity can automatically sort the observation targets and detect them in order. AEGIS will continue to be applied in the "Mars 2020" exploration mission, and so on.

In terms of health prediction and fault diagnosis: NASA can use artificial intelligence software Livingstone Version 2 (LV2) to automatically detect and diagnose simulated faults in the instruments and systems of the Earth Observation One (EO-1) satellite, and so on.

In terms of human–computer interaction and learning control: DL and other complex algorithms are difficult to apply to space-based systems, but they have made outstanding achievements in ground-based systems supporting space missions. Google and NASA have successfully discovered two planets in two distant stellar systems through the ML technology of a large amount of data from the Kepler telescope, and the number of planets in one star has reached 8, which is called "the second solar system." According to the AIDA (artificial intelligence data analysis) project funded by the European Vision 2020 framework, using various data from our solar system, ESA, and NASA are developing an intelligent system that will read and process data from space, aiming to achieve new discoveries, reveal anomalies, and identify structures.

In terms of intelligent management and autonomous operation: NASA is studying how to make the communication network more efficient and reliable by using cognitive radio. Cognitive radio selects "white noise" areas in the communication frequency band and uses them to transmit data, making maximum use of the available limited telecommunication frequency band, and minimizing the delay time, etc.

In terms of space intelligence design and manufacturing: AI technology is also used *in the design process of satellites.* NASA's Evolutionary software uses AI to quickly design a tiny advanced space antenna. The software starts with a random antenna design and is optimized through the evolution process. The computer system took about 10 hours to complete the initial antenna design process. Although the method adopted in this project is based on genetic algorithm, it is now completely possible to use complex algorithms such as DL in the satellite design process to perform tasks such as weight reduction optimization, system layout, mechanical design, etc.

3.2.2 European Space Agency (ESA)

In terms of human–computer interaction and learning manipulation: ESA's Advanced Concept Team (ACT) recently studied evolutionary computing to write computer code in a way that considers all evolution. Among them, better results are kept, and bad results are discarded. The whole process is like biological evolution. One application is to calculate the orbits of planets [12].

In terms of spatial intelligent planning and decision-making: ACT has also conducted research using ML *in the fields* of guidance, navigation, and control. They study the use of ant colony robots that share information in the network: if it is beneficial for a robot to learn mobility from experience, then the whole group will learn this, called the ant colony algorithm. Other examples of AI activities supported by ACT include the Community Science Mobile App, which will improve the autonomous capability of space probes and optimize the star tracking system.

In terms of human–computer interaction and learning control: ESA has begun to use many AI and ML in its space missions. For example, walkers can bypass obstacles by finding their own way in the "unknown" field. The intelligent data transmission software on the Mars probe eliminates human scheduling errors;

otherwise, valuable data will be lost. In addition, ESA has gained rich experience in using AI to process large amounts of data to extract meaningful information.

3.2.3 Other Countries and Regions

In terms of human–computer interaction and learning control: The German Aerospace Center recently launched an AI assistant to support the astronauts' daily tasks on the International Space Station. CIMON (Crew Interactive Mobile Company), completely controlled by voice, can communicate with astronauts by watching, speaking, and listening.

In terms of space intelligent perception and recognition: Japan Space Agency has developed an intelligent system and is currently taking test photos of the Japanese module KIBO on the International Space Station. JAXA's Int-Ball can run independently and take photos and videos. Its development aims to promote the autonomy of off-vehicle and in-vehicle tests and, at the same time, seek to obtain the robot technology necessary for future exploration missions [13, 14].

The summary of the application of intelligent technology for major satellites is shown in Table 3.2.

3.3 Development Trend Analysis

At present, the common problems of satellites are slow response to emergencies, poor timeliness of in-orbit data, strong dependence on the ground, and difficulty in adapting to changes in space environment. The development goal of artificial intelligence technology in the future aerospace field is to realize the *weak intelligence of satellites*, so that they have the ability to intelligent perception, recognition and understanding, autonomous management and reconstruction, autonomous operation and management of large systems, group intelligence, and autonomous coordination, autonomous planning, and intelligent control, to achieve the final autonomous reasoning and thinking of unmanned vehicles, and to solve the problems encountered in emergencies [15–17].

1) *Launch AI development strategy in aerospace field*

The current international space development situation urgently requires scientific research institutions to closely track the technological frontier and consolidate the technological reserves. Future space model missions pose higher demands and challenges for the intelligent level of space vehicle control systems. "Artificial intelligence in the field of space" is imperative. Standing at the forefront of technological development, we need to seize the opportunity to take off and realize the leapfrog and even subversive development of space technology.

Table 3.2 Summary of main satellite application intelligent technology.

Serial number	Technical direction	Time	Task name/type	Manufacturer	Introduction to AI application
1.	Spatial intelligent perception and recognition	Launch in 2009	Kepler Mission Kepler space telescope	NASA Google	*Google and NASA jointly used machine learning to discover the "second solar system": in* 2017, NASA and Google announced that with the help of machine learning, a large number of data collected by the Kepler Space Telescope (14 billion data points) will be processed through automated software and manual analysis, and *15000 marked Kepler signals will be used to train the machine learning model to distinguish planetary signals (with an accuracy of up to 96%),* The eighth planet Kepler-90i, known as the second solar system, has been found orbiting Kepler-90. In addition, AI also discovered another star, Kepler-80g, the smallest planet in the Kepler-80 galaxy.
2.		Deployed in 2000	The deployed satellite platforms of SensorWeb Experiences sensor network system include EO-1, HyspIRI, Terra&Aqua, MidSTAR	NASA	*NASA used AI to discover Iceland's volcanic eruption: the* SensorWeb sensor network system was designed and implemented by JPL Lab. The system is used in the global monitoring program to monitor and study volcanoes, fires, floods, etc. The system uses sensor networks linked by software and the Internet to achieve autonomous satellite observation and response capability
3.		Launched in 2001	BIRD satellite Bispectral infrared remote sensing detection satellite	DLR German Aerospace Center ISRO Indian Space Research Organization	The BIRD satellite is equipped with infrared remote sensing detector based on neural network classifier, which can realize real-time and rapid disaster monitoring and early warning

4.	Launched in 2013	IPEX satellite "intelligent payload experiment" technology test satellite	NASA	IPEX satellite is designed to pave the way for the technology of HyspIRI, a business intelligent satellite. The preliminary test requires a powerful target recognition algorithm based on statistical principles
5.		Spatial intelligent planning and decision-making		
		Mars 2020 Mars rover 2020	NASA	The design goal of Mars 2020 vision system working mode is *fully intelligent autonomous navigation*. Based on the artificial intelligence of self-learning perception, it is planned to collect 20 Mars core and soil samples in 1.25 Mars years (28 Earth months).

Three important directions for NASA to use AI technology in future Mars exploration missions (such as Mars cave exploration, etc.) are *automatic driving of Mars rovers*; the *rover independently identifies, classifies, and collects samples of Martian rock core, soil, and other chemical components*; the *intelligent router will allow the rover to independently adjust its route according to the schedule*. |
| 6. | Launched in 1998 | Deep Space 1 "Deep Space One" detector | NASA | "Deep Space One" has realized the flight test of fully autonomous optical navigation technology for the first time. It does not rely on the ground measurement and control system, only uses the imaging system carried on the satellite to obtain the images of asteroids and stars, quickly calculates the position of the deep space probe, and forms the orbit correction command |

(Continued)

Table 3.2 (Continued)

Serial number	Technical direction	Time	Task name/type	Manufacturer	Introduction to AI application
7.		Launched in 2005	"Deep Impact" detector	NASA	The "Deep Impact" detector uses the navigation camera to capture the comet image, extract the information of the target comet, use the autonomous navigation method used in the "Deep Space One," and combine the attitude determination system to give the detector attitude, and independently complete the position and velocity state estimation of the detector
8.		Launched in 2003	"Falcon" asteroid detector	JAXA Japan Aerospace Exploration Agency	The "Falcon" probe launched by Japan's JAXA uses optical navigation to achieve the approach of asteroids and finally successfully achieve asteroid sampling
9.		Launch in 2011	MSL Curiosity Rover Curiosity rover	NASA	The intelligent perception decision system of Mars rover MSL, based on OASIS (CLEAR execution) system, *combines computer vision with machine learning and can find new scientific objectives of albedo characteristics* through training
10.		Launched in 2003	Opportunity and Spirit rovers	NASA	It is equipped with "Hybrid Active Activity Planning Generator" (MAPGEN) for task planning. MAPGEN can automatically generate planning and scheduling schemes based on high-level scientific task instructions on the ground, perform hypothesis testing, support planning editing, analyze resource usage, perform constraint execution and maintenance, and form low-level action instructions, thus simplifying the burden of task control personnel and increasing scientific returns. "Opportunity" and "Curiosity" rovers have the initial intelligent perception ability

11.	Health prediction and fault diagnosis	Launch in 2009	TacSat-3 Tactical Satellite-3	NASA AFRL US Air Force Research Laboratory	TacSat-3 studies the trust problem of *automatic reasoning system* through the launch management test, and the automatic reasoning system is mainly used for fault detection and diagnosis. This satellite is a verification example for NASA and AFRL to use as *intelligent space systems*
12.		Deployed in 2004	Earth Observing-1 Earth Observer 1 satellite	NASA	NASA deployed Livingstone 2 (L2) health management software on EO-1 satellite in 2004. L2 has hardware diagnosis, multiple assumptions, reverse tracking, single-event transient pulse diagnosis, code and model decoupling, 17 diagnostic scenarios and long-term space operation diagnosis, etc.
13.		Launched in 2003	Mars Express Mars Express spacecraft carries Beagle-2 (Beagle 2 lander)	ESA UKSA	The detector is equipped with the artificial intelligence tool MEXAR2 detection data scheduling system, which can intelligently predict on-board data packets may be lost due to memory conflicts, optimize the data download plan and generate the commands required for actual transmission. Compared with traditional methods, MEXAR2 greatly reduces the workload of the task planning team, basically eliminates the loss of stored data packets, and generates a feasible download plan
14.		Launched in 2003	"Falcon" asteroid detector	JAXA Japan Aerospace Exploration Agency	JAXA applied ISACS-DOC fault diagnosis system on the "Falcon" asteroid detector. However, at present, the autonomous fault handling system has problems such as insufficient capacity to handle emergencies

(Continued)

Table 3.2 (Continued)

Serial number	Technical direction	Time	Task name/type	Manufacturer	Introduction to AI application
15.	Human–computer interaction and learning control	Deployment in 2011	Robonaut-2 robot astronaut 2 is deployed at the International Space Station	NASA GE	The Robonaut Program began in 1996 and is divided into three stages. First, the Robonaut will be limited to work at a fixed position in the International Space Station; it will then be allowed to move freely within the space station; finally, carry out the final extravehicular activities. According to the sensory input received, the robot astronauts can make their *own decisions about the* next action arrangement. The *intelligent perception* of Robonaut-2 robot in the United States is reflected in the communication between astronauts and the autonomous completion of fine operations
16.		Deployed in 2006	SPHERES Synchronous position maintenance, participation and repositioning experimental satellite/robot is deployed at the International Space Station	NASA DARPA Defense Advanced Research Projects Agency MIT	SPHERES robots use ultrasonic and infrared detection technology to perform basic navigation tasks in certain areas of the International Space Station. The acoustic signal device is used to triangulate the environment, and the microphone on the satellite surface will obtain the acoustic signal to accurately locate it.

The intelligent SPHERES robot is *equipped with Android operating system and can be controlled by mobile phone* |

No.				
17.	Deployment in 2018	CIMON Simon is deployed at the International Space Station	DLR German Aerospace Center ESA IBM Airbus	As an *AI assistant*, CIMON can communicate with astronauts by watching, speaking and listening. An additional camera is dedicated to face recognition; the cameras on both sides mainly record video information or meet other computer-generated functions (such as augmented reality); ultrasonic sensor measures the distance of collision detection; a directional microphone for speech recognition; the core voice processing unit of AI is Watson system developed by IBM; AI autonomous navigation module is provided by Airbus, mainly used for motion planning and object recognition. CIMON cannot learn independently, and must be trained by people
18.	Launch 3A in 2016 Launch 3B in 2018	Sentinel-3A Sentinel 3A satellite Sentinel 3B satellite	ESA Thales Alenia Space Act – Italian Thales Alenia Aerospace	Terez Alenia Aerospace put forward the Tomorrow Factory Initiative, aiming at developing innovative technology centered on workers, integrating the most advanced technologies and supporting facilities such as additive manufacturing, *augmented reality, and collaborative robots*, which has been continuously promoted for 2yr
19.	Deployment in 2013	Kirobo is deployed at the International Space Station	JAXA The University of Tokyo TOYOTA	The Kirobo robot can *communicate with astronauts, detect gestures and recognize astronauts' emotions at* the same time, and react. When it senses that astronauts are in bad mood, it will take the initiative to comfort them
20.	/	/	American DSC Company US Army	*According to the brainwave training neural network model, the* United States DSC Company and the United States Army use brainwave training machine shooting. The research of the laboratory shows that if the computer can learn the EEG data related to target location, it can narrow the gap between the computer and human in this area

(Continued)

Table 3.2 (Continued)

Serial number	Technical direction	Time	Task name/type	Manufacturer	Introduction to AI application
21.	Intelligent management and autonomous operation	Deployed in 2012	Scan Testbed space communication and navigation test platform is deployed at the International Space Station	NASA	The SCaN Testbed inter-communication and navigation test bed is an on-orbit laboratory built by NASA for the International Space Station, which is used to demonstrate and verify the new communication technology, networking technology and navigation technology of radio (SDR) to promote the development of the new generation of space communication technology. *Using AI and machine learning, satellites can seamlessly control the space communication system and make real-time decisions without instructions*
22.		Launched in 2000	Earth Observing-1 Earth Observer 1 satellite	NASA	*The ASE software system carried by EO-1 can conduct scientific exploration independently and respond to scientific events on the earth independently; LV2 software system can automatically detect and diagnose analog faults in satellite instruments and systems*
23.	Spatial intelligent design and manufacturing	Launched in 2006	Space Technology 5/ST-5	NASA	The three ST-5 satellites are equipped with computer collective design space antennas, which are *tiny advanced space antennas quickly designed by combining 80 personal computers based on artificial intelligence technology*
24.		First launch in 1996	A2100 geosynchronous spacecraft series A2100 satellite series platform	LMT Lockheed Martin	Loma uses 3D printing to manufacture titanium satellite parts through *intelligent manufacturing* technology. In the manufacturing process, titanium is heated, and then almost any shape is created by stacking layer by layer. The material waste is minimal, and the manufacturing cycle is greatly reduced

2) *Strengthen the development of intelligent satellite system*

We will continue to strengthen the application of new AI technologies, develop high-performance landmark satellites, breakthrough on-board processing capabilities, intelligence and other technologies, and improve satellite capabilities and usability. Explore intelligent control, intelligent information acquisition, intelligent communication network and other new technologies, aim at earth observation, communication, manned, deep space exploration, space attack and defense and other fields and decision-making needs, and vigorously develop original intelligent satellite systems.

   (i) AI technology will become the most important means to meet the development and production needs of large and giant constellations.

  (ii) The integration of AI technology and on-orbit operation technology may become an effective solution for the development and on-orbit operation of large space systems.

 (iii) Satellite manufacturing, testing, and commissioning will develop from automation to autonomy and intelligence.

3) Focus on the development of AI algorithms and key supporting devices

In the aerospace field, most satellites are in the stage of weak AI or transiting to strong AI. Aerospace engineering is a complex and huge system engineering, and artificial intelligence technology cannot be directly applied. Limited by factors such as algorithm research and hardware support (insufficient onboard computing capacity), AI technology must be changed and upgraded according to the needs of the project. In terms of hardware, such as chips with high-speed processing and very low energy consumption in space environment, artificial intelligence image detectors, etc. In terms of software, develop AI algorithms applicable to various fields of aerospace [15, 18, 19].

Human–computer cooperation will become the main direction of AI applications. With the complementary nature between human beings and AI systems, the *collaborative interaction* between human beings and AI systems will become the main direction of AI applications. Although fully autonomous AI systems will play an important role in underwater or deep space exploration and other applications, AI systems cannot completely replace humans in the short term in disaster recovery, medical diagnosis, and other applications. Human-machine cooperation can be divided into three types: *joint execution, auxiliary execution, and alternative execution.* As shown in Table 3.3.

Table 3.3 Three modes of human–computer cooperation.

Serial number	Name	Human–computer cooperation mode
1.	*Joint implementation*	AI system positioning: perform peripheral tasks supporting human decision-makers; typical applications: short-term or long-term memory retrieval and prediction tasks
2.	*Auxiliary execution*	AI system positioning: when human beings need help, the artificial intelligence system performs complex monitoring functions; typical applications: ground proximity alarm system, decision-making, and automatic medical diagnosis in aircraft
3.	*Alternative execution*	AI system positioning: AI systems perform tasks with very limited capabilities for humans; typical applications: complex mathematical operations, dynamic system control, and guidance in controversial operating environment, automatic system control in hazardous or toxic environment, nuclear reactor control room, and other rapid response systems

References

1 Antsaklis, P.J. (1994). Defining Intelligence Control – Report of task force on intelligent control. *IEEE Control System Magazine* 14 (3): 4–5, 58–66.

2 Gundy-Burlet, K., Krishnakumar, K., Soloway, D., and Kaneshige, J. (2002). *Intelligent Control Approaches for Aircraft Applications*. National Aeronautics & Space Administration Ames Research.

3 EPSS (2018). The Age of Artificial Intelligence-Towards a European Strategy for Human-Centric Machines. EPSC Strategic Notes (29): 1–14.

4 Huet, C. (2018). Artificial Intelligence Strategy for Europe. European Commission.

5 ASGARD (2018). Artificial Intelligence – A strategy for European startups. ROLAND BERGER GMBH.

6 Vincent, J. (2019). AI systems should be accountable, explainable, and unbiased, says EU. https://www.theverge.com/2019/4/8/18300149/eu-artificial-intelligence-ai-ethical-guidelines-recommendations. Accessed 8 April 2019.

7 European Space Agency (2019). Whitepaper. Block chain and earth observation.

8 Scott, B., Heumann, S., and Lorenz, P. (2018). Artificial intelligence and foreign policy. Political Science.

9 Authority of the House of Lords (2017). AI in the UK: ready, willing and able. Report of Session 2017-19.

10 Marr, B. (2018). Government uses artificial intelligence to identify welfare and state benefits fraud. https://www.forbes.com/sites/bernardmarr/2018/10/29/how-the-uk-government-uses-artificial-intelligence-to-identify-welfare-and-state-benefits-fraud/. Accessed 29 October 2018.

11 Hall, D.W. and Pesenti, J. (2017). Growing the artificial intelligence industry in the UK. Computer Science.

12 Zhukov, B., Lorenz, E., Oertel, D. et al. (2006). Spaceborne detection and characterization of fires during the bi-spectral infrared detection (BIRD) experimental small satellite mission (2001–2004). *Remote Sensing of Environment* 100 (1): 29–51.

13 Fesq, L. (2003). *Rover Autonomy System Validation*. MSL Focused Technology Task.

14 Biesiadecki, J.J. and Maimone, M.W. (2006). The Mars Exploration Rover surface mobility flight software: driving ambition. In: *2006 IEEE Aerospace Conference*, Big Sky, MT, USA (04–11 March 2006). IEEE. 0-7803-9546-8/06.

15 Nayak, P., Kurien, J., Dorais, G. et al. (1999). Validating the DS-1 remote agent experiment. In: *Proceedings of the 5th International Symposium on Artificial Intelligence, Robotics and Automation in Space*, 349. Paris: European Space Agency.

16 Ahlstrom, T., Curtis, A., Diftler, M. et al. (2013). Robonaut 2 on the International Space Station: status update and preparations for IVA mobility. In: *AIAA SPACE 2013 Conference and Exposition*, 5340.

17 Cornelius, R. (2016). International Space Station (ISS) payload autonomous operations past, present and future. In: *SpaceOps Conferences*, Daejeon, Korea (16–20 May 2016). SpaceOps.

18 Baird, D. (2017). NASA Explores Artificial Intelligence for Space Communications. www.nasa.gov/scan (accessed 09 December 2017).

4

Basic Knowledge of AI Technology

4.1 The Concepts and Characteristics of Machine Learning and Deep Learning

At the beginning of 1950, AI pursued developing machines that could have intelligence like human beings. The research community called this "strong AI." Later, expert systems emerged, which applied AI technology in specific fields to inject new vitality into the development of AI. However, it also brought problems such as difficult transplantation and high cost. Since 1980, machine learning has become the mainstream of AI research. It studies how computers simulate or realize human learning behavior to obtain new knowledge or skills, and reorganizes existing knowledge structure to continuously improve its performance. Around 2000, computer scientists added multilayer perceptron to build a deep learning model based on neural network research and successfully solved many problems in the fields of image recognition, speech recognition, and natural language processing. In recent years, driven by technology giants such as IBM, cognitive computing has flourished. Through learning and understanding unstructured data such as languages, images, and videos, we can better obtain knowledge from massive complex data and make more accurate decisions. Machine learning is one of the core issues in the field of artificial intelligence. The theoretical achievements have been applied to various fields of artificial intelligence. Machine learning algorithms divide things into different categories according to their characteristics through pattern recognition systems. Through the selection and optimization of the recognition algorithm, it has a stronger classification ability [1].

Explore relevant concepts in the field of artificial intelligence. Machine learning, deep learning, and reinforcement learning are the core driving force for its rapid development. Artificial intelligence was the first to appear and was once highly valued; after that, machine learning, a small subset of AI, has developed rapidly; in

Intelligent Satellite Design and Implementation, First Edition. Jianjun Zhang and Jing Li.
© 2024 The Institute of Electrical and Electronics Engineers, Inc. Published 2024 by John Wiley & Sons, Inc.

the past 10 years, another branch in the field of machine learning, deep learning, has received widespread attention, has made an unprecedented impact, and has become the core of driving the AI explosion. The following will briefly describe the concepts and characteristics of machine learning and deep learning [2].

4.1.1 Artificial Intelligence: Give Human Intelligence to Machines

As early as the meeting in the summer of 1956, the pioneers of artificial intelligence dreamed of using the computer that had just appeared to construct complex machines with the same essential characteristics as human intelligence. This is what we now call "General AI." This omnipotent machine has all our senses (even more than people), and all our rationality can think like us.

People always see such machines in movies: friendly, like C-3PO in Star Wars, Evil, such as the Terminator. Strong AI still exists only in movies and science fiction. The reason is not hard to understand. We have not been able to achieve them, at least not yet [3].

What we can achieve at present is generally called "Narrow AI." Weak AI is a technology that can perform specific tasks like or even better than people. For example, image classification on Pinterest; or Facebook's face recognition.

4.1.2 Machine Learning: A Method to Realize Artificial Intelligence

Unlike the traditional method of directly writing a computer program to complete a certain algorithm directly from input to output, the machine learning method is a method in that the computer uses the existing data (experience) to obtain a certain model (the law of lateness), and uses this model to predict the future (whether late or not). Therefore, to some extent, machine learning is similar to the experience process of human thinking, but it can consider more situations and perform more complex calculations. In fact, one of the main purposes of machine learning is to transform the process of human thinking and inductive experience into the process of computer processing and calculating data to obtain models. The computer model can solve many flexible and complex problems in a way similar to humans. The process of machine learning is consistent with the way of human thinking and follows the basic process of "discovering rules through data and making predictions" (Figure 4.1).

Machine learning comes directly from the early field of artificial intelligence. Traditional algorithms include decision tree learning, deductive logical programming, clustering, reinforcement learning, Bayesian network, etc. In terms of learning methods, machine learning algorithms can be divided into supervised (such as classification problems), unsupervised learning (such as clustering problems), semi-supervised learning, ensemble learning, and reinforcement learning. The early machine learning methods were limited by the computing power of

Figure 4.1 Schematic diagram of machine learning.

machines and even could not achieve weak AI. With the passage of time, the development of computer ability has changed everything [4].

4.1.3 Deep Learning: A Technology to Realize Machine Learning

Deep learning is not an independent learning method but also uses supervised and unsupervised learning methods to train neural networks. However, due to the rapid development of this field in recent years, some unique learning methods have been proposed one after another (such as residual networks), so more and more people regard it as a learning method alone.

Initial deep learning is a learning process that uses a deep neural network to solve feature representation. The deep neural network itself is not a new concept, which originates from the artificial neural network. An artificial neural network is an important algorithm in early machine learning, which has been through ups and downs for decades. The principle of neural networks is inspired by the physiological structure of our brain – interconnected neurons. But unlike a neuron in the brain that can connect any neuron within a certain distance, the artificial neural network has discrete layers, connections, and data propagation directions.

The deep neural network can be roughly understood as a neural network structure containing multiple hidden layers. In order to improve the training effect of deep neural networks, people have made corresponding adjustments to the connection method and activation function of neurons. In fact, there were many ideas in the early years, but the final effect was not satisfactory because of insufficient training data and poor computing ability at that time [5].

Deep learning has achieved various tasks in a way that makes it possible for all machine auxiliary functions. Driverless cars, preventive medical care, and even better movie recommendations are all near or about to be realized. The reasons are closely related to the following factors:

1) The deep neural network needs a lot of data for training. If the network depth is too shallow, the recognition ability is often not as good as the general shallow model, such as SVM or boosting; if you do it deeply, you need a lot of data for training; otherwise overfitting will be inevitable in the learning process. Since 2006, the rapid development of the Internet has produced a large number of data for training and learning, namely the so-called big data era.

2) The second is computing power. Deep network requires high computing power of computer. A large number of repeatable and parallelizable calculations are required. At that time, when the CPU had only a single core, and the computing power was relatively low, it was impossible to carry out deep network training. With the growth of GPU computing capacity, it is possible to combine deep networks with big data training.

3) Finally, review the development process of a neural network, from a single-layer neural network (perceptron) to a two-layer neural network including a hidden layer, and then to multilayer deep neural network, which has experienced "three ups and three downs," as shown in Figure 4.2. This is between the ups and downs, with the insistence of many scientists, such as Lecun and Hinton, and finally ushered in the dawn of deep neural networks taking the mainstream.

4.1.4   Reinforcement Learning: Self-evolution Mechanism of Learning Feedback

Reinforcement learning is inspired by the fact that organisms can effectively adapt to the environment, interact with the environment through the mechanism of trial and error, and learn the best strategy by maximizing cumulative rewards.

The reinforcement learning system is composed of four basic parts: state s, action a, state transition probability $P_{s,s'}^{a}$, and reward signal r. Policy $P_i: S \rightarrow A$ is defined as the mapping from state space to action space. The agent selects action a according to the policy P_i in the current state S, executes the action, and transfers to the next state s' with probability $P_{s,s'}^{a}$. At the same time, the reward r received from the environment feedback is received. The goal of reinforcement learning is to maximize the cumulative reward by adjusting the strategy. Usually, the value function is used to estimate the pros and cons of a certain strategy P_i.

In the process of practical application, it is often combined with deep learning and reinforcement learning applications to detect and identify effective information through deep learning and then form systematic feedback evolution through reinforcement learning. As early as 2015, Google's artificial intelligence research team, DeepMind, published two remarkable research results in the past two years: the deep reinforcement learning algorithm based on Atari video game and the first game of computer Go No. 1. These efforts broke the shackles of the traditional academic design of human-like intelligent learning algorithms, and combined the deep learning (DL) with the ability of perception and the reinforcement learning (RL) with the ability of decision-making to form the deep reinforcement learning (DRL) algorithm. The principle framework is shown in Figure 4.2. The excellent performance of these algorithms is far beyond people's imagination and has greatly shocked academic and social circles [6].

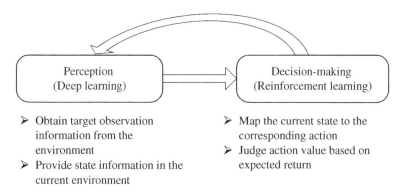

Figure 4.2 Deep Reinforcement Learning Framework.

The two articles of the Deep Intelligence Team in Nature magazine make DRL become the focus of advanced AI. The deep Q-network (DQN) proposed in January 2015 has made a breakthrough in Atari video games. DQN simulates the process of human players playing games, directly uses the game screen as information input, and the game score as the reinforcement signal of learning researchers tested the algorithm after training convergence and found that its score in 49 video games was higher than that of human advanced players. On this basis, the Deep Intelligence Team further put forward the computer Go Chuyi in January 2016. The algorithm combines the DRL method with Monte Carlo tree search (MCTS), which greatly reduces the computation of the search process and improves the accuracy of chess game estimation. In the match with the European Go champion Fan Hui, Chuyi won a 5-0 victory. In March 2016, Chuyi played a world-famous game with Lee Sedol, the world's top chess player, and finally won the game 4 : 1. This also indicates that DRL, as a new machine learning algorithm, has been able to compete with human beings in complex chess games [7].

4.1.5 Transfer Learning: A New Machine Learning Method that Uses Existing Knowledge to Solve Problems in Different but Related Fields

Transfer learning relaxes two basic assumptions in traditional machine learning: (i) the training samples used for learning and the new test samples meet the conditions of independent and identical distribution. (ii) There must be enough available training samples to learn a good classification model. The purpose is to transfer the existing knowledge to solve the learning problem that there is only a small amount of labeled sample data or even no data in the target domain.

In traditional classification learning, in order to ensure the accuracy and high reliability of the classification model obtained by training, there are two basic assumptions. However, in practical applications, we find that these two conditions cannot be met. First, with the passage of time, the previously available tagged sample data may become unavailable, creating a semantic and distribution gap with the distribution of new test samples. For example, stock data is very time-sensitive data. The model learned from the training samples of last month cannot predict the new samples of this month very well. In addition, sample data with labels are often scarce and difficult to obtain. In the field of Web data mining, new data is emerging constantly, and the existing training samples are not enough to train a reliable classification model while labeling a large number of samples is very time-consuming and laborious, and because human subjective factors are prone to errors, this has caused another important problem in machine learning, how to use a small number of labeled training samples or source domain data, Establish a reliable model to predict the target domain data (source domain data and target domain data may not have the same data distribution). It is pointed out that data classification should first solve the problem of sample sampling in training sets, and how to select representative sample sets as training sets is an important issue worth studying [8].

AI technology, from the traditional simple adaptive feedback algorithm to the simple machine learning method, to the increasingly practical and popular DRL and migration methods in recent years, has become increasingly mature in practical applications, which has a disruptive impact on traditional industries, especially the big data-driven industries. In view of this, Japan, the United States, and Western European countries have formulated relevant research plans. For example, the US Department of Defense has invested US $400 million to develop an eight-year research plan by the Defense Advanced Research Projects Agency (DAPRA) and set up corresponding organizations and steering committees. At the same time, the Office of Naval Research (ONR) and the Office of Air Force Scientific Research (AFOSR) have also invested huge amounts of money in the research of neural networks. DARPA believes that neural network "seems to be the only hope to solve machine intelligence," and that "this is a more important technology than atomic bomb engineering." National Science Foundation (NSF), National Aeronautics and Space Administration (NASA), and other government agencies also attach great importance to the development of neural networks, and they support many research topics in different forms [9].

4.2 Key Technologies of AI

4.2.1 Classification of Key AI Technologies

Key AI technologies and their definitions are listed in Table 4.1.

Table 4.1 Key AI technologies and their definitions.

Serial number	Technical name	Connotation
1.	Machine learning	Machine Learning is an interdisciplinary subject involving statistics, system identification, approximation theory, neural network, optimization theory, computer science, brain science, and many other fields. It is the core of artificial intelligence technology to study how computers simulate or realize human learning behavior in order to obtain new knowledge or skills, reorganize existing knowledge structure to continuously improve its performance. Data-based machine learning is one of the most important methods in modern intelligent technology. It is studied to find laws from observed data (samples) and use these laws to predict future data or unobservable data. According to its learning mode, machine learning can be divided into supervised learning, unsupervised learning, and reinforcement learning. According to the learning method, it can be divided into traditional machine learning and deep learning
2.	Computer vision	Computer vision is the science of using computers to imitate human visual system, which enables computers to have the ability to extract, process, understand, and analyze images and image sequences similar to human beings. Automatic driving, robot, intelligent medical, and other fields need to extract and process information from visual signals through computer vision technology. Recently, with the development of deep learning, preprocessing, feature extraction, and algorithm processing are gradually integrated to form end-to-end AI algorithm technology. According to the problems solved, computer vision can be divided into five categories: computational imaging, image understanding, three-dimensional vision, dynamic vision, and video coding and decoding
3.	Virtual reality/ augmented reality	Virtual reality (VR)/augmented reality (AR) is a new audio-visual technology with computers as its core. In combination with relevant science and technology, a digital environment that is highly similar to the real environment in terms of vision, hearing, touch, etc., is generated within a certain range. Users interact with objects in the digital environment with the necessary equipment, interact with each other, and obtain a feeling and experience similar to the real environment through display devices, tracking and positioning devices, tactile interaction devices, data acquisition devices, special chips, etc.

(Continued)

Table 4.1 (Continued)

Serial number	Technical name	Connotation
4.	Human-computer interaction	Human-computer interaction mainly studies the information exchange between people and computers, mainly including human–computer and computer-to-human information exchange. It is an important peripheral technology in the field of artificial intelligence. Human-computer interaction is a comprehensive discipline closely related to cognitive psychology, ergonomics, multimedia technology, virtual reality technology, etc. Including voice interaction, emotional interaction, somatosensory interaction, brain-computer interaction, etc.
5.	Knowledge map	Knowledge map is essentially a structured semantic knowledge base. Different entities are connected with each other through relationships, forming a network knowledge structure. In the knowledge map, each node represents the "entity" of the real world, and each edge represents the "relationship" between entities. Generally speaking, the knowledge map is a network of relationships obtained by connecting all different kinds of information, providing the ability to analyze problems from the perspective of "relationships."
6.	Biometric recognition	Biometric identification technology refers to the technology of identifying and authenticating individual identity through individual physiological or behavioral characteristics. The content of biometric recognition technology is very extensive, including fingerprint, palm print, face, iris, finger vein, voice print, gait, and other biological features. Its recognition process involves image processing, computer vision, speech recognition, machine learning, and other technologies. At present, as an important intelligent identity authentication technology, biometric identification has been widely used in finance, public security, education, transportation, and other fields
7.	Natural language processing	Natural language processing is an important direction in the field of computer science and artificial intelligence. Research on various theories and methods that can achieve effective communication between people and computers using natural language involves many fields, mainly including machine translation, machine reading comprehension, and question answering system

4.2.2 Technical Development Trend Analysis

1) The AI basic platform is open source, and the open-source DL software framework will usher in a golden age of development.

   With the continuous in-depth integration of AI applications in production and life, the demand for the function and performance of the in-depth learning software framework will gradually erupt, resulting in many related tools and open source in-depth learning software frameworks, and lowering the threshold of AI application deployment. The open-source DL framework allows developers to directly use the developed DL tools, reduce secondary development, improve efficiency, promote close cooperation and communication in the industry, and have a great impact on the field of DL. Google, Baidu, and other industrial giants have also established an industrial ecosystem through open-source technology to seize the commanding heights of the industry. Through the open-source of technology platforms, expand the scale of technology, integration of technology and applications, and effective layout of the entire industrial chain of AI, more software and hardware enterprises will participate in the open-source ecosystem in the future [10].

2) The development of AI technology from special AI to general AI.

   The development of AI is mainly focused on dedicated intelligence, which has domain limitations. With the development of science and technology, various fields are integrated and interact with each other. A universal intelligence with a wide range, high integration, and strong adaptability is needed to upgrade from auxiliary decision-making tools to professional solutions. General AI has the ability to perform general intelligent behaviors. It can connect AI with human characteristics such as perception, knowledge, consciousness, and intuition, reduce dependence on domain knowledge, and improve the universality of processing tasks. This will be the future development direction of AI. In the future, AI will cover a wide range of fields and eliminate application barriers between various fields.

3) The research and application of transfer learning will become an important direction.

   Because transfer learning focuses on the research of knowledge transfer, parameter transfer, and other technologies in DL, it can effectively improve the reusability of DL models and also provides a method for the interpretation of DL models, which can provide theoretical tools for the reliability and inexplicability of DL algorithm models, so it will become an important direction for its research and application in the future development of artificial intelligence.

4.3 Machine Learning

Machine learning is an important research branch in the field of artificial intelligence. The main content of its research is to make use of computer programs so that machines with processors and computing functions can improve the performance of problem processing with the increase of experience. At present, machine learning theory has been widely applied to intelligent video surveillance, biometrics, unmanned driving, and other fields [11].

Machine learning can be basically divided into four stages. The first stage: from the 1950s to the mid-1960s, the system improved the system parameters and its performance through continuous learning of input and output feedback. The second stage: from the mid-1960s to the mid-1970s, this stage is mainly about the study of the system structure, such as explaining and describing the internal structure of the machine through logical structure or graph structure. The third stage: from the mid-1970s to the mid-1980s, machine learning has made great progress by studying learning strategies and methods to improve learning efficiency and introducing knowledge databases. Since 1986, with the introduction of neural networks and the need for artificial intelligence, people have studied the connection mechanism of machine learning [12].

At present, the research in machine learning can be divided into four categories: unsupervised learning, supervised learning, semi-supervised learning, and RL.

4.3.1 Unsupervised Learning

Unsupervised learning is a self-learning classification method. It learns from unmarked training samples to discover hidden structural relationships between unknown data. Unsupervised learning and kernel density estimation methods are very similar. The commonly used unsupervised learning includes association rule learning and cluster learning.

Unsupervised learning is a classification method of autonomous learning without providing labels manually; that is, by learning training samples that are not labeled in advance, the inherent structural relationship in the data is mined. Since the samples input to the classifier are completely unmarked, the classifier will not get the corresponding feedback information to evaluate the learning results after processing these data. This is also one aspect of unsupervised learning that is different from supervised machine learning. Most unsupervised learning algorithms are based on clustering. Unsupervised learning can be used not only as a process to find the internal distribution structure of data but also as a preprocessing process of other classification learning tasks, that is, first cluster the sample data by unsupervised learning and then sign the classification results, and then use supervised learning to train the classification model, in this way, automatic classification can be realized when there is new sample data input. Commonly used unsupervised learning algorithms include K-means algorithm, EM algorithm, etc.

4.3.2 Supervise Learning

In the process of machine learning, supervised learning can increase human participation and provide instruction labels. This feature allows human participation, so that the algorithm can reduce the error of the machine itself. This type of learning is used more in classification and prediction. Supervised learning obtains a general learning model by learning the training sample data set with a given tag. Once there is new data outside the sample data set that needs similar analysis, the learning model obtained in the training stage can be used to predict the results of the new data. Generally, supervised learning is divided into three steps: the first step is to mark the sample, the second step is to train, and the third step is to estimate the probability of the model. The general process is as follows:

a) Input the feature vector and sample category mark of the sample.
b) During training, the prediction results are compared with the actual marking of the training samples by analyzing the feature vectors of the samples.
c) Adjust the prediction model until the accuracy of the prediction model is consistent with the expected accuracy. Common algorithms include:
   1) K-nearest neighbor algorithm
   K-nearest neighbor algorithm is a basic and simple classification and regression algorithm. Its basic approach is to determine the k-nearest neighbor training instance points of the input instance points for the given training instance points and input instance points, and then use the number of classes of the k training instance points to predict the class of the input instance points. The advantage of k-nearest neighbor algorithm is that its implementation mechanism is relatively simple. It does not need to display the training process. It just collects the data, saves the sample data after specifying the label, and then processes it after inputting the test sample. The disadvantages of this algorithm are shown in two aspects. The first is the selection of k value, which cannot be measured by specific criteria. The second is that k is generally selected as an odd number. Because when k is an even number, there are two or more classes in the category of the nearest k neighbor points to the test sample, then the category cannot be determined [13].
   2) Naive Bayesian algorithm
   Naive Bayesian method is a classification method based on Bayesian theorem and independent assumption of characteristic conditions. In the algorithm, it uses probability and statistics knowledge to make classification decisions.

In many cases, the effect of naive Bayesian classification algorithm can be comparable to that of a decision tree and neural network classification algorithm. The algorithm can not only be applied to large databases but also has the advantages of simple method, fast speed, and high classification accuracy. In the Bayesian theorem, it is assumed that the effect of an attribute value on a given class has nothing to do with the value of other attributes. However, this assumption is often not true in practice. Therefore, it will affect the classification accuracy and may reduce its accuracy. Therefore, there are many Bayesian classification algorithms that reduce the assumption of independence, such as the tree-enhanced naive Bayesian algorithm. Compared with other algorithms, the naive Bayesian algorithm has the following advantages: simple implementation and calculation and can achieve better classification results when the classification objects are text data and numerical data. The disadvantage is that its classification accuracy depends on the degree of independence of each attribute. The algorithm assumes that each attribute is independent of each other. If the dataset can meet this assumption of independence, the accuracy of classification is relatively high [14].

3) Support vector machine

Support vector machine is a supervised learning method that applies statistical learning theory to it. SVM suppresses the over-fitting of the function by controlling the interval measure of the hyperplane, thus obtaining the linear optimal decision function. Due to the introduction of the kernel function, the sample vector can be mapped to the high-dimensional feature space, and then the optimal classification surface can be constructed in the high-dimensional space to obtain the linear optimal decision function. At the same time, because of the introduction of kernel function, the problem of dimension has also been solved ingeniously. The main advantages of support vector machines are: insensitive to high-dimensional and sparse data, better capture the internal characteristics of data, and high accuracy. The disadvantage is that for nonlinear problems, the kernel function is the largest variable that affects the application effect of support vector machine, so the selection of kernel function is very important but also very difficult. If the selection of kernel function is inappropriate, it means that the sample is mapped to an inappropriate space, which may lead to poor performance [15].

4) Decision tree algorithm

Decision tree is a machine learning algorithm that uses a tree structure to make decisions. A decision tree contains these parts: a root node, several internal nodes, and several leaf nodes. By adopting a recursive idea, branch from the root node to the internal node and finally reach the leaf node. The process of branching from the root node of the tree structure to the internal node is to compare the attribute values of the sample according to a certain splitting principle and determine whether the node will continue to branch or terminate according to the characteristics or size of the attribute values. The node that will not split at last is called the leaf node, which is the final category to be divided. From the root node at the beginning to the leaf node at the end, the path formed is equivalent to an expression selection criterion, and the entire decision tree is composed of a set of criteria for selecting expressions. In 1986, Quinlan proposed the famous ID3 algorithm. This decision tree learning algorithm is based on information gain. However, because the information gain criterion tends to select the attribute with more values, in fact, in most cases, the multi-value attribute is not necessarily the optimal attribute. In addition, ID3 algorithm is generally effective for samples with small data sets, and it is also sensitive to noise. When the training data set becomes larger, the decision tree may change. Based on ID3 algorithm, Quinlan then proposed C4.5 algorithm. C4.5 algorithm selects the optimal partition attribute based on the information gain rate. This algorithm does not directly select the candidate partition attribute with the largest gain rate but uses a heuristic method. First, find out the attribute with the information gain higher than the average level from the candidate partition attribute, and then select the one with the highest gain rate. For the processing needs of large-scale data sets, some improved algorithms were proposed later, among which CART (Classification and Regression Tree) and random forest algorithm are quite representative algorithms. CART decision tree algorithm is based on the "Gini index" to optimize the division of attributes. Random forest is a forest (combined classifier) composed of a series of decision trees (subclassifiers) by random methods [16].

4.3.3 Deep Learning

DL is a new research direction in the field of machine learning. In recent years, it has made breakthroughs in speech recognition, computer vision, and other applications. Its motivation is to establish a model to simulate the neural connection structure of the human brain. When processing images, sounds, and text signals, the data features are described through multiple transformation stages, and then the interpretation of the data is given.

The reason why DL is called "depth" is that compared with "shallow learning" methods such as support vector machine, boosting method, and maximum entropy method, the model learned by DL has more layers of nonlinear operations. Shallow learning relies on manual experience to extract sample features, and network model learning results in single-layer features without hierarchical structure. DL transforms the feature representation of samples in the original space into a new feature space through layer-by-layer feature transformation of the original signal, and automatically learns to obtain hierarchical feature representation, which is more conducive to classification or feature visualization. Another theoretical motivation of the DL theory is that if a function can be expressed in a concise form with a k-layer structure, then the expression with a k-1 layer structure may require an exponential number of parameters (relative to the input signal), and the generalization ability is insufficient [17–22].

4.3.3.1 Feedforward Depth Network

Feedforward depth network feedforward neural network is one of the original artificial neural network models. In this network, information flows in only one direction, from the input unit to the output unit, through one or more hidden layers, and there is no closed loop in the network. Typical feedforward neural networks include multilayer perceptron and convolutional neural networks. The perceptron proposed by F. Rosenblatt is the simplest single layer.

Forward artificial neural network, but then M. Minsky et al. proved that single-layer perceptron cannot solve linear indivisible problems (such as XOR operation). This conclusion led the research field of artificial neural networks to a low tide until researchers realized that multilayer perceptron can solve linear indivisible problems, and the research on the combination of back-propagation algorithm and neural network made the research of neural network become a hot spot again. However, due to the disadvantages of the traditional back-propagation algorithm, such as slow convergence speed, the need for many labeled training data, and easy to fall into local optimum, the effect of multilayer perceptron is not very ideal.

4.3.3.2 Convolution Neural Network

Convolution neural network is a generalization of neurocognitive machine. Convolution neural network is a trainable multilayer network structure composed of multiple single-layer convolution neural networks. Each single-layer convolution neural network includes three stages: convolution, nonlinear transformation, and down-sampling. The down-sampling stage is not necessary for each layer. The input and output of each layer is a feature map composed of a set of vectors (the original input signal of the first layer can be regarded as a high-dimensional feature map with high sparsity).

In recent years, convolution neural network has achieved great success in large-scale image feature representation and classification. The AlphaGo AI Go program, which defeated Li Shishi by 4 : 1 in the famous Go human-computer game in April 2016, adopted CNN + Monte Carlo search tree algorithm. The convolutional neural network was first proposed by le Cun et al. in 1998 for the recognition of handwritten character images. The input of the network is the original two-dimensional image. After several convolution layers and full connection layers, the prediction probability of the output image in each category is obtained. Each convolution layer contains three operations: convolution, nonlinear activation function, and maximum pooling. The advantages of using convolution operation are as follows:

1) The two-dimensional convolution template can better mine the local relationship between adjacent pixels and the two-dimensional structure of the image.
2) Compared with the fully connected structure of the general neural network, the convolutional network greatly reduces the number of network parameters through weight sharing, making it feasible to train large-scale networks.
3) The convolution operation is robust to the translation, rotation, and scaling of the image.

After obtaining the convolution response characteristic diagram, it is usually necessary to go through a nonlinear activation function to obtain the activation response diagram, such as sigmoid, tanh, and Re LU functions. Next, apply a maximum pooling or average pooling operation on the activation function response graph. In this operation, the feature map is first divided into several spatial areas with uniform grid, which can have overlapping parts, and then the average or maximum value of each image area is taken as the output. In addition, in maximum pooling, it is usually necessary to record the position of the output maximum value. Because the convolution neural network has many parameters, it is easy to over-fit and affect the final test performance. Hinton et al. proposed an optimization technique called "drop out," which can prevent over-fitting by randomly ignoring half of the feature points in each training iteration and has achieved certain results. Wan et al. further extended this idea. In the training of the full connection layer, a subset randomly selected from the connection weight of the network is set to 0 at each iteration, so that each network update is targeted at different network structures, further improving the generalization of the model. In addition, there are some simple and effective engineering techniques, such as momentum method, weight decay, and data enhancement.

4.3.3.3 Cyclic Neural Network

In the fully connected DNN and CNN networks, the signals of each layer of neurons can only be transmitted to the upper layer, and the processing of samples is independent at each time. Therefore, this kind of neural network cannot model the changes in time series. For example, the time sequence of the appearance of samples is very important for natural language processing, speech recognition, handwriting recognition, and other applications. To meet this demand, there is another neural network structure – cyclic neural network. The output of neurons in RNN can directly affect itself at the next time stamp, that is, the input of layer i neurons at time t. In addition to the output of layer i-1 neurons at time t-1, it also includes its own input at time t. This achieves the purpose of modeling time series.

4.3.3.4 Bidirectional Deep Network

The bidirectional network is formed by the superposition of multiple encoder layers and decoder layers. Each layer may be a separate encoding process or decoding process, or may contain both encoding and decoding processes. The structure of bidirectional network combines two kinds of single-layer network structures, i.e. encoder and decoder, while the learning of bidirectional network combines the training methods of feedforward network and feedback network, which usually includes two parts of single-layer network pretraining and layer-by-layer reverse iteration error. The pretraining of single-layer network mostly uses greedy algorithm to train each layer of network. After each layer of network structure has been pretrained, the weight of the whole network structure is fine-tuned through reverse iteration error. The pretraining of single-layer network is the reconstruction process of encoding and decoding the input signal, which is similar to the training method of feedback network. The weight fine-tuning based on the reverse iteration error is similar to the feedforward network training method.

4.4 Natural Language Processing

Natural language processing is the study of various theories and methods to achieve effective communication between people and computers through natural language. The research on natural language processing began in the late 1940s and the early 1950s and has developed for more than 60 years. It has made considerable progress and formed a relatively mature theoretical system. Natural language processing methods have been widely used in speech recognition, text translation, big data processing, artificial intelligence, and other fields [23].

Through the analysis of speech sounds, the input speech is divided into phonemes, and phonemes are recognized into corresponding morphemes according to the rules and then combined into words according to the rules between morphemes. After having a word, determine the meaning of the word through part of speech analysis and grammar analysis, and then you can get a complete sentence. According to a certain context, you can define the meaning of the sentence more accurately. Further, we can analyze the meaning of the whole conversation through the scene and then combine common sense knowledge to understand its purpose [24].

Of course, from the current common application scenarios of natural language processing, the complete processing process will not be used in most cases. For example, in the application of voice input, the process of 1–6 is generally used. The application of written language translation generally uses a process of 4–7.

4.4.1 Introduction to Natural Language Processing Methods

The methods of natural language processing are mainly divided into two directions.

One is to sum up the feature rules of language and establish the rule base through feature engineering. Then the computer analyzes and processes the natural language according to the characteristics in the rule base. The advantage of this method is that the processing process is relatively direct, the processing of complex sentences can be relatively flexible, and the demand for the original corpus is low, which can be put into use quickly. However, the coverage of language knowledge is relatively low, and it is often necessary to update the feature library when encountering new problems to ensure normal use. And when the feature library becomes larger and larger, a more complex design is needed to avoid conflicts between features [25].

The other is to establish a mathematical model through the relevant theories of statistics, and the computer continuously analyzes and calculates the original corpus to optimize the parameters in the mathematical model. Finally, this mathematical model is used to process natural language. This way gives the computer the ability of self-learning. Through the continuous training of the parameters of the model through the corpus, it can improve the processing ability of the model, expand its language knowledge coverage, and improve the processing ability. However, this method also has some defects. For example, when it comes to complex semantics, it is often difficult to get correct results. Or when the original corpus used for training is insufficient, the model parameters will be inaccurate, and the final processing results will also be inaccurate. In addition, due to the need to establish mathematical models through computers according to statistical theory, the ability of developers is relatively high [26].

With the establishment of large-scale corpora of various languages all over the world, statistics-based natural language processing methods have gradually become mainstream, and many mathematical models based on statistical methods have emerged. However, in recent years, the research of statistical methods has also encountered bottlenecks. The main reason is that the statistical method is to build models based on the frequency characteristics of language, which is contrary to people's common sense. Therefore, further research found that more satisfactory results can be obtained by combining some methods of characteristic laws when applying statistical methods to build models. For example, building the original corpus into a structured tree-based form, and so on.

4.4.2 Introduction of Natural Language Processing Model

After a long period of development, natural language processing has formed some commonly used processing models. This section will briefly introduce some processing models:

1) *Regular expression model*: Regular expressions are a great innovation in the computer field. With the help of regular expressions, we can obtain the specific strings we need from the character stream in an extremely concise way. Because of its simplicity and efficiency, regular expressions have been widely used in various fields of computer. For natural language processing, regular expressions can help us obtain the required characters and phrases through specific rules.

2) *N-ary grammar model*: This is a widely used language processing tool. It calculates the conditional probability through the dominant condition and then calculates the processing result according to the maximum likelihood estimation method. This method needs a lot of training to get the available probability parameters, so it is highly dependent on the corpus used for training.

3) *Hidden Markov model*: Hidden Markov model is a statistical model, which is used to describe stochastic processes with unknown parameters. By establishing a dynamic Bayesian network and then using forward and backward algorithms, the probability distribution of each node on the network is continuously deduced according to the known state. Through a continuous learning process, hidden Markov model can infer unknown parameters from known parameters. At present, hidden Markov model has been successfully applied to the field of speech recognition [27].

4.5 Knowledge Engineering

As a new stage of AI development, knowledge science and knowledge engineering have broad application prospects in the knowledge society. Based on the analysis of some problems existing in the current research of AI, it is proposed to promote the development of AI from the theoretical and practical levels by strengthening the research of knowledge science and developing knowledge engineering. Knowledge engineering is a subject that takes knowledge as the processing object and borrows the engineering idea to study how to use the principles, methods, and technologies of artificial intelligence to design, construct, and maintain knowledge-based systems. The purpose of knowledge engineering is to develop intelligent systems based on research knowledge. Therefore, knowledge acquisition, knowledge representation and knowledge application constitute the three major elements of knowledge engineering [28].

4.5.1 Expert System

The core of knowledge engineering is expert system. Expert system is an intelligent computer program that uses knowledge and reasoning steps to solve complex problems that only experts can solve. That is, any computer program whose problem-solving ability has reached the level of human experts in the same field can be called an expert system. In a certain professional field, the knowledge-based computer system may be equivalent to the role of human experts in this field [29].

4.5.1.1 Development of Expert System

The first expert system project, DENDRAL Chemical Molecular Structure Analysis System, was studied at Stanford University in 1965 and was successfully studied in 1968. The system can infer the molecular structure of unknown organic compounds from the mass spectrometer data. It is a heuristic system that can quickly eliminate the impossible molecular structure and avoid the exponential expansion of search space. By generating all possible molecular structures, he can even find out the structures that human experts might miss. After commercialization, the system has been widely used in the world, providing chemical structure explanations for hundreds of international users every day.

The expert system has attracted the attention of all countries in the world for its performance and practicability. The United States, Japan, Europe and other developed countries have invested a lot of scientists in research and have developed many high-level expert systems, which can be like human experts.

It also solves many complicated and difficult problems in the application field. MYCIN system, like an experienced infectious disease doctor, can diagnose and treat infectious diseases for patients. American Digital Equipment Company can save tens of millions of dollars annually by using the computer configuration expert system XCON. Expert systems can create great economic benefits, and even some small microcomputer-based expert systems can also create great benefits. After investigating the application performance of expert systems in many countries and regions in the world, Professor Feigenbaum, the authority of artificial intelligence and the father of knowledge engineering, concluded that the use of expert systems could save a lot of money and almost all expert systems can at least improve the working efficiency of people by 10 times, some by 100 times, or even 300 times [30].

At present, the application fields of expert systems have penetrated many fields such as mathematics, physics, chemistry, biology, agriculture, geology, meteorology, transportation, metallurgy, chemical engineering, machinery, politics, military, law, space technology, environmental science, information management system, finance, and information superhighway, and its application has penetrated almost all walks of life.

With the improvement of the overall level of computer and science and technology, the research of new generation expert systems such as distributed expert systems and collaborative expert systems has also developed rapidly [31].

4.5.1.2 Expert System and Its Functions

Explanatory expert system: Its function is to explain the actual meaning of information and data through the analysis of information and data. Typical interpretative expert systems include signal understanding and chemical structure interpretation. For example, the DENDRAL system, the language understanding system HEARSAY, which interprets the molecular structure of compounds from mass spectrometer data. It is to find out the interpretation consistent with the known data and consistent with the objective law.

Diagnostic expert system: Its function is to find out the faults and problems existing in the processed objects according to the input information. There are various diagnostic expert systems in medical, mechanical, and electronic fields. For example, DART, a computer hardware fault diagnosis system, is to detect and find possible faults by dealing with the functions of various components within the object and the relationship between them.

Debugging expert system: Mainly to provide the troubleshooting scheme of confirmed faults. It mainly includes computer-aided debugging expert systems for electronic equipment and mechanical equipment. For example, TIMM/TUNER, the auxiliary debugging system of the VAX/VMS computer system, selects the best scheme from multiple error correction schemes according to the characteristics of processing objects and faults.

Maintenance expert system: Its main function is to formulate and implement a plan to correct a certain type of fault. For example, the computer network maintenance expert system, the telephone cable maintenance expert system ACE system, etc., are based on the characteristics of the fault and error correction methods of the object system to formulate a reasonable maintenance plan.

Characteristics of students, and organizes the advantages of the knowledge set that needs to be learned with the appropriate teaching methods and lesson plans, which is used to teach and guide students, diagnose and deal with students' mistakes in learning. For example, CAI is a computer-aided teaching system, and EXCHECK is a good teaching system of logic and set theory.

Predictive expert system: Its function is to analyze and speculate on the future evolution and development based on the past and present situation of the processing object. Typical applications include weather forecast, economic development forecast, traffic forecast, etc. For example, the market forecast system of products, weather forecast system, etc.

Planning expert system: Its main function is to find a certain action sequence or step that can achieve a given goal. Planning expert system has been used for robot action planning, transportation scheduling, engineering project demonstration and planning, and production operation planning.

Design expert system: Its function is to form the required pattern or graphic description according to the given requirements. Typical applications include circuit design and mechanical design. For example, the overall structure and configuration system XCON of VAX computer, automatic programming system PSI, etc.

Monitoring expert system: Its function is to complete real-time monitoring tasks, continuously monitor the behavior of the system, object or process, and compare the monitored behavior with the behavior it should have, to find abnormal conditions and send out an alarm. Typical applications include air traffic control monitoring and nuclear power plant safety monitoring [32].

Control-type expert system: Its function is to adaptively manage all behaviors of a controlled object or group to meet the expected requirements. It is generally used for real-time control tasks. For example, battlefield auxiliary combat command system, vehicle transmission command system, production process control and air traffic control.

Various types of expert systems can be interrelated. Some expert systems usually need to complete several types of tasks at the same time. For example, MYCIN is a diagnostic and debugging expert system.

At present, most expert systems are analytical expert systems, and the problems solved are classification problems. The basic operation of classification problem-solving is called interpretation operation. When the input data and corresponding output data are given, it is required to give whether the system is abnormal and the reason for the abnormality. The interpretation operation is mainly to identify the operation, which is the input and output of the object system. When the input data and specific object system are given, it is required to explain what kind of output is expected, and the interpretation operation is the prediction operation. When the specific object system and its output are given, the problem to be solved is to determine the required input, and the interpretation operation is the control operation. The knowledge base of the analytical expert system consists of three parts: data (evidence) set, hypothesis (solution) set, and heuristic knowledge that links data and hypothesis. Their possible combination constitutes the state space or problem-solving space, and the search and solution are carried out in this limited space. The key of developing an expert system is to express and apply expert knowledge. At present, expert systems mainly use rule-based knowledge representation and reasoning technology. Because domain knowledge is more imprecise or uncertain, uncertain knowledge representation and knowledge reasoning are important topics in the development and research of expert systems [33].

Programming language is the most basic tool for developing expert systems. LISP language and PROLOG language are two artificial intelligence languages that can conveniently represent knowledge and reasoning technology. C++, with an object-oriented style, and traditional languages such as C and PASCAL are also common languages for constructing expert systems.

4.5.1.3 Structure of Expert System

The structure of expert system refers to the construction method and organization form of each component of expert system. The appropriateness of the system structure selection is closely related to the applicability and effectiveness of the expert system. In the book, a rule-based expert system is selected. The knowledge base contains domain knowledge encoded in the form of rules to solve problems. The system is mainly composed of the following parts:

a) *User interface*: the communication mechanism for interaction between users and expert systems. Generally, graphical user interface is used to facilitate the use of ordinary users.
b) *Interpreter*: the interpretation system compiles the reasoning according to the interpretation principle and transmits it to the user in a way that the user can understand.
c) *Working memory*: facts in the global database used by rules.
d) *Reasoning machine*: decide which rules are consistent with the facts or objectives, then grant the corresponding priority to the rules, and finally execute the rule with the highest priority to complete the reasoning.
e) *Agenda*: a rule priority table created by the inference engine, which is used to complete the fact or target in the rule matching working memory according to the rule priority table.
f) The knowledge acquisition machine establishes an automatic knowledge input mode for users to replace knowledge engineers to acquire knowledge.

4.5.2 Data Mining

1) *Definition*
   Data Mining, commonly known as knowledge discovery in databases in the field of artificial intelligence, is an extraordinary process of obtaining valuable, effective, and ultimately understandable information from a large amount of data. In short, data mining is to extract or mine knowledge from huge data.
2) *Function*
   Data mining aims to discover hidden and meaningful knowledge from the database. Its functions can be divided into two categories: first, descriptive data mining: generating new and unusual information based on available data sets; second, predictive data mining: based on the existing data sets, the system model described by the known data sets is generated. These two types generally include the following functions.

(i) *Concept description*

Conceptual description is used to describe and summarize the connotation of a certain type of object and its related characteristics. Conceptual description can be realized by the following two methods: data characterization and data differentiation. The former is used to describe the common characteristics of certain objects, such as starting from the characteristics of different services of China Mobile and China Unicom, and finding potential customers in the future. The latter describes the difference between different types of objects, such as bank card fraudsters and nonfraud fraudsters, and compares the characteristics of the two types of bank card holders.

(ii) *Association analysis*

Association analysis is to find the meaningful relationship, correlation, causal structure, and frequent patterns between different itemsets from a large amount of data. If two or more data item values are repeated, and the probability is high, there will be some correlation between them. You can establish association rules between these data items. The purpose of association analysis is to find out the hidden association rules in the database. For example, customers who buy mobile phones will also buy a certain type of battery, which is an association rule.

In the process of mining association rules, there are usually two important indicators: support and confidence. Only rules with high support and confidence are more valuable and can be used as reference rules.

3) *Classification and prediction*

Classification is to find a group of typical characteristic functions or models that can describe the data set, to be able to classify and identify the category or attribution of unknown data. For example, bank card users are divided into high, medium, and low-risk groups, and customers are divided into defined groups. Among them, the classification model or function can be learned from a group of training sample data (its category and attribution are known) through a classification mining algorithm, and it has many forms of representation, such as decision tree, classification rules, neural network, or mathematical formula.

The prediction is to use historical data to find out the rules, establish experimental models, and use this model to predict the types and characteristics of future data. For example, predict which customers will cancel the cooperation with the company next year or which credit card users will apply for other value-added services.

4) *Cluster analysis*

Clustering is also called unsupervised learning. Clustering aims to divide data into a series of meaningful subsets according to certain rules. In the same cluster, the distance between individuals is small, while in different clusters, the distance between individuals is large. For example, according to the price fluctuations of different types of funds, different types of funds can be divided into different categories, which types of funds can be divided into, and the characteristics of each category. This may be very important information for buyers, especially for fund investors. Of course, in addition to classifying samples, clustering can also complete the mining of outliers, such as applying it to fraud detection. Clustering and classification are different. Cluster analysis is a method to complete information clustering based on information similarity without giving classification. Classification is guided learning that needs to define classification categories and training samples first.

5) *Outlier analysis*

Outliers refer to some data contained in the database that is inconsistent with the general behavior of the model or data. Most data mining methods generally treat outliers as anomalies or noises and discard them. In some applications, it is very necessary to find outliers from the database, such as commercial fraud, abnormal income or criminal behavior, and carry out outlier analysis. Therefore, finding and analyzing outlier data is a very meaningful data mining task.

6) *Evolutionary analysis*

Evolution analysis is to model and analyze the changing laws and development trends of data objects that change over time. It mainly includes time series analysis, other series, or periodic pattern matching and analysis based on similarity data. For example, 70% of people who have bought mobile phones will choose to buy new batteries after two years.

7) *Implementation steps*

Data mining is the first and core step of knowledge discovery (KDD) in databases.

a) *Determine the exploration object*

A clear definition of the problem to be explored and a clear understanding of the purpose of data mining is the key step in data mining. The result of mining is unpredictable, but the problems to be explored should be predictable. For data mining, data mining is blind and cannot be successful.

b) *Data preparation*

Data preparation can be divided into three steps: data selection, data preprocessing, and data conversion. The purpose of data selection is to search all internal or external data information related to the exploration object and select data suitable for data mining applications. The purpose of data preprocessing is to study the quality of data, prepare for further analysis, and determine the type of mining operation to be carried out. The purpose of data transformation is to transform data into an analysis model, which is built for mining algorithms. The key to the success of data mining is to establish an analysis model that is truly suitable for mining algorithms.

c) *Data mining*

To mine the data obtained after data preparation, all other work can be completed automatically except for selecting the appropriate mining algorithm.

d) *Result analysis*

Explain and evaluate the results. The analysis method used should be determined by the data mining operation, usually using visualization technology.

e) *Assimilation of knowledge*

Integrate the knowledge obtained from the analysis into the organizational structure of the business information system. Decision-makers can adjust their competitive strategies according to the results of data mining and the actual situation.

The data mining process needs repeated cycles to achieve the expected results [34].

8) *Classification of data mining technology*

Data mining can be classified according to the adopted technology. The most common methods are:

a) *Rule induction*: That is, to find valuable rules. For example, rule mining on association is to find the internal relationship between some items in the database. This was first used in supermarkets and now has been used by more enterprises. By using rule induction, we can find potential customers, analyze new business models, and so on, to help the development of business and contribute to the improvement of the competitiveness of enterprises [35].

b) *Decision tree method*: Mainly uses the relevant rules generated by classification to carry out a synthesis. It mainly uses data entropy to analyze and search for information, then establishes a tree-shaped trunk structure according to relevant rules, and then establishes relevant branches according to different fields, and finally forms a tree-shaped structure rule through this repeated construction. ID3 method is a very classical decision tree method.

c) *Artificial neural network*: The neural structure of the human brain can be simulated by modern computer technology. Through thousands of times of continuous learning, a nonlinear prediction model can be formed to classify massive data, and then the correlation analysis of data can be carried out by the clustering method.

d) *Genetic algorithm*: This is an algorithm that simulates the biological evolution process. It was first proposed by Holland in the 1970s. As the name implies, he uses the data according to the relevant methods of genetics. In genetics, there are four obvious characteristics that genes can be combined, crossed, mutated, and selected. Genetic algorithm to process data based on these four methods. Just like gene processing, everyone is coded and given a corresponding fitness value. To adapt to this genetic algorithm, data mining technology needs to undergo relevant processing.

e) *Fuzzy technology*: use the relevant fuzzy theory in mathematics to make fuzzy predictions and analysis of the actual problems encountered, and then make relevant identification and decision. The fuzziness exists objectively. The higher the complexity of the system, the stronger the fuzziness. The biggest characteristic of fuzzy set theory is which one is like this one. Li Deyi proposed a new model called the cloud model, which has become a cloud theory after continuous development and improvement. This theory is very valuable in data mining and provides new methods for data mining.

f) *Rough set method*: It was developed by Polish logician Z. Pawlak in 1982. It has been widely used in machine learning in recent years. This method mainly studies some fuzzy problems in the information system that cannot be determined. It is based on some ideas of equivalence classes. The main method of research is to scramble all the attribute values of information, then divide it again according to certain rules, and finally simplify some information to obtain the desired information [36].

g) *Visualization*: It refers to expressing abstract things in a more intuitive way for the next step of processing. Visualization technology mainly includes data, model, and process visualization. Data visualization mainly includes histograms, box-shoulder diagrams, and scatter diagrams. In fact, the method of visualization is closely related to data mining. We use the shape of the tree to represent the decision tree. Visualization is to describe the whole process of knowledge discovery with some relatively intuitive things, such as pictures.

References

1 Diaz, D., R-Moreno, M.D., Cesta, A. et al. (2011). Applying AI action scheduling to ESA's space robotics. In: *11th Symposium on Advanced Space Technologies in Robotics and Automation Proceedings*, Moreno, 5B.

2 Vassev, E. and Hinchey, M. (2013). On the autonomy requirements for space missions. In: *16th IEEE International Symposium on Object/Component/Service-Oriented Real-Time Distributed Computing*. IEEE.

3 Jonsson, A.K. (2007). Spacecraft autonomy: intelligent software to increase crew, spacecraft and robotics autonomy. In: *2007 Conference and Exhibit*, Rohnert Park, California (7–10 May 2007), 2007–2791. AIAA.

4 Shafto, M. and Sierhuis, M. (2010). AI space odyssey. *IEEE Intelligent Systems* 25 (5): 16–19.

5 Clancey, W.J., Sierhuis, M., Alena, R. et al. (2007). Automating CapCom using mobile agents and robotic assistants. In: *1st Space Exploration Conference: Continuing the Voyage of Discovery* (19 January 2005), 2005–2659. AIAA.

6 Schuster, A.J. (2007). *Intelligent Computing Everywhere*. Springer-Verlag London Limited.

7 Garret, R. (2012). The Remote Agent Experiment: Debugging Code from 60 Million Miles Away. YouTube.com. Google Tech Talks. Slides (14 February 2012).

8 Chang, M.A., Bresina, J., Charest, L. et al. (2004). MAPGEN: mixed-initiative planning and scheduling for the mars exploration rover mission. *IEEE Intelligent Systems* 8–12.

9 Sherwood, R., Chien, S., Tran, D. et al. (2005). Intelligent systems in space: the EO-1 autonomous sciencecraft, AIAA Paper-6917. *Proceedings of the Infotech@Aerospace Conference*, Arlington, VA (26–29 September 2005): Infotech@Aerospace.

10 Clement, B.J., Barrett, A.C., Rabideau, G.R., and Durfee, E.H. (2001). Using abstraction to coordinate multiple robotic spacecraft. In: *Proceedings of the 2001 Intelligent Robots and Systems Conference*, Maui, HI (November 2001).

11 Wang, Y.Q., Ma, L., and Tian, Y. (2011). State-of-the-art of ship detection and recognition in optical remotely sensed imagery. *Acta Automatica Sinica* 37 (9): 1029–1037.

12 Zhang, S., Du, Z., Zhang, L. et al. (2016). Cambrion-X: An Accelerator for Sparse Nural Networks. In: *2016 49th Annual IEEE ACM International Symposium on Microarchitecture (MICRO)*, Taipei, Taiwan, 1–12. IEEE.

13 Han, S., Kang, J., Mao, H., and Yiming, H. (2017). ESE: Efficient Speech Recognition Engine with Sparse LSTM on FPGA. In: *ACM/SIGDA International Symposium on Field-Programmable Gate Arrays*, Monterey, California, USA, 75–84. ACM/SIGDA.

14 Jouppi, C.Y., Patil, N., Patterson, D. et al. (2017). In-datacenter performance analysis of a tensor processing unit. In: *The 44th International Symposium on Computer Architecture (ISCA)*, Toronto, Canada, 1–17. ISCA.

15 Suda, N., Chandra, V., Dasika, G. et al. (2016). Throughput-optimized OpenCL-based FPGA accelerator for large-scale convolutional neural. In: *ACM/SIGDA International Symposium on Field-Programmable Gate Arrays*, Monterey, California, USA, 16–25. ACM/SIGDA.

16 Liu, Z., Dou, Y., and Jiang, J. (2017). Throughput-optimized FPGA accelerator for deep convolutional neural networks. *ACM Transactions on Reconfigurable Technology and Systems* 10 (3): 1–23.

17 Dicecco, R., Lacey, G., Vasiljevic, J. et al. (2017). Caffeinated FPGAs: FPGA framework for convolutional neural networks. In: *International Conference on Field-Programmable Technology*, 265–268. Melbourne, Australia: IEEE.

18 Zhou, Y. and Jiang, J. (2016). An FPGA-based accelerator implementation for deep convolutional neural networks. In: *The International Conference on Computer Science and Network Technology (ICCSNT)*, 829–832. Changchun, China: IEEE.

19 Zhang, C., Li, P., Sun, G. et al. (2015). Optimizing FPGA-based accelerator design for deep convolutional neural networks. In: *ACM/SIGDA International Symposium On Field-Programmable Gate Arrays*, Monterey, California, USA, 161–170. ACM/SIGDA.

20 Zhang, C., Wu, D., Sun, J. et al. (2016). Energy-efficient CNN implementation on a deeply pipelined FPGA cluster. In: *ACM International Symposium on Low Power Electronics and Design (ISLPED)*, San Francisco Airport, CA, USA, 326–331. ACM/SIGDA.

21 Chen, Y.H., Krishna, T., Emer, J.S., and Sze, V. (2017). Eyeriss: an energy-efficient reconfigurable accelerator for deep convolutional neural networks. *IEEE Journal of Solid-State Circuits* 52 (1): 127–138.

22 Reagen, B., Whatmough, P., Adolf, R. et al. (2016). Minerva: enabling low-power, high-accuracy deep neural network accelerators. In: *Proceedings of ACM/IEEE International Symposium on Computer Architecture* (18–22 June 2016). IEEE.

23 Simonyan, K. and Zisserman, A. (2015). Very deep convolutional networks for large-scale image recognition. *International Conference on Learning Representations* (7–9 May 2015). San Diego, CA, USA, ICLR

24 Girshick, R., Donahue, J., Darrell, T., and Malik, J. (2015). Region-based convolutional networks for accurate object detection and segmentation. *IEEE Transactions on Pattern Analysis and Machine Intelligence 38* (1): 142–158.

25 Xiao-qiang, Z., Boli, X., and Gang-yao, K. (2015). A ship target discrimination method based on change detection in SAR imagery. *Journal of Electronics & Information Technology* 37 (1): 63–70.

26 Xiao-bo, L., Wen-fang, S., and Li, L. (2015). Ocean moving ship detection method for remote sensing satellite in geostationary orbit. *Journal of Electronics & Information Technology* 37 (8): 162–1867.

27 Zhicheng, G., Huiyi, Z., and Jihong, P. (2013). A method for ship detection based on neighborhood characteristics in remote sensing image. *Journal of Shenzhen University Science and Engineering* 30 (6): 584–591.

28 Yu, J.-y., Dan, H., Wang, L.-y. et al. (2016). Real-time on-board ship targets detection method for optical remote sensing satellite. In: *2016 IEEE 13th International Conference on Signal Processing (ICSP)*, 204–208. IEEE.

29 Yu, J.-y., Dan, H., Wang, L.-y. et al. (2016). Real-time on-board ship targets detection method based on multi-scale salience enhancement for remote sensing satellite. In: *2016 IEEE 13th International Conference on Signal Processing*, 217–221. IEEE.

30 Li, W.J., Zhao, H.P., and Shang, S.N. (2017). Onboard ship saliency detection algorithm based on multi-scale fractal dimension. *Journal of Image and Graphics* 22 (10): 1447–1454.

31 Huang, J., Jiang, Z.G., Zhang, H.P. et al. (2017). Ship object detection in remote sensing images using convolutional neural networks. *Journal of Beijing University of Aeronautics and Astronautics* 43 (9): 1841–1848.

32 Wei, X., Chen Yanton, P.I.A.O., and Yongjie, W.S. (2017). Target fast matching recognition of on-board system based on Jilin-1 satellite image. *Optics and Precision Engineering* 25 (1): 255–261.

33 Dahl, G.E., Yu, D., Deng, L. et al. (2012). Context-dependent pre-trained deep netural networks for large-vocabulary speech recognition. *IEEE Transactions on Audio, Speech, and Language Processing* 20 (1): 30–42.

34 Gooddfellow, I.J., Shlens, J., and Szegedy, C. (2015). Explaining and harnessing adversarial example EB/OL]. *arXiv preprint* arXiv: 1412.6572v3stat.ML].

35 Silver, D., Huang, A., Maddison, C. et al. (2016). Mastering the game of Go with deep neural networks and tree search. *Nature* 529 (7587): 484–489.

36 Cesta, A., Cortellessa, G., Denis, M. et al. (2007). MEXA2: AI solves mission planner problems. *IEEE Intelligent Systems* 22 (4): 12–19.

5

AI Requirements for Satellite System

5.1 Demand Requirements for AI Technology in Satellite System

From the perspective of problem orientation, the demand for satellite systems for artificial intelligence mainly includes the following aspects: the need to improve the management ability through intelligent means. When the task is complex, and the environment is bad, the satellite needs to have certain active adaptability. In addition, when the artificial auxiliary response is slow, especially for some remote exploration missions, such as Mars exploration missions, it takes about 20 minutes for a signal to go back and forth. If it is only controlled by the ground, it is not timely, so the satellite needs to have its own independent capabilities, such as independent judgment, independent decision-making, independent planning, and independent execution. There is also a demand for artificial intelligence technology, and so on [1, 2].

Although the United States actively applied AI technology to remote sensing satellites, space telescopes, Mars probes, and other satellite systems as early as the beginning of this century, most of the technologies used are non-AI technologies in a strict sense such as logical reasoning and computer vision, as well as a small number of symbolic intelligence and data-driven machine learning and other representative methods derived from the first wave of AI, the application of the technology of the second and third wave of artificial intelligence represented by neural network and deep learning is still in the trial stage. However, it is well known that these technologies have made epoch-making achievements in the civil field, and their great application value in the aerospace field cannot be ignored.

In short, from the perspective of system and technology, the application requirements of satellite systems for AI technology are roughly in the following five

Intelligent Satellite Design and Implementation, First Edition. Jianjun Zhang and Jing Li.
© 2024 The Institute of Electrical and Electronics Engineers, Inc. Published 2024 by John Wiley & Sons, Inc.

directions: spatial intelligent perception, spatial intelligent decision-making and control, spatial cluster intelligence, spatial intelligent interaction, and spatial intelligent design [3].

5.1.1 Space Intelligent Perception Requirements

The basic premise for space exploration and transformation activities of space vehicles is to have a comprehensive and full understanding of their space environment, including physical space, relative/absolute time, mission objects, partners, etc. In the face of explosive data input presented by the continuous accumulation of time and space, as well as a variety of fuzzy or indirect information, the perception of space vehicles needs to have a high level of intelligence, in order to extract real and effective information from a large number of data, obtain comprehensive and direct information from fuzzy and indirect data, and provide strong support for the next step of decision-making [4].

Space intelligent perception refers to the perception and comprehensive judgment of the space vehicle on the target, environment, and its own state by intelligent means.

The goal of space intelligent perception is to provide enough high-precision and real-time information for the satellite system to carry out ground situation awareness, on-orbit service, deep space exploration, and other tasks based on the particularity of the space environment for the intelligent, autonomous, and real-time requirements of the satellite system in the process of carrying out space missions. By introducing machine learning methods represented by deep learning and combining intelligent means such as data mining, the satellite system can realize situation awareness and high-precision measurement of the space environment, space targets, and ground targets.

The application requirements of satellite systems for space intelligence perception include: remote sensing satellite ground, space target recognition, situation analysis, three-dimensional reconstruction of deep space probe environment, detection target recognition and processing, space robot kinematics, dynamic parameter identification, health monitoring, etc.

The key technologies involved in spatial intelligent perception include deep learning, 3D reconstruction, SLAM based on deep learning, system identification, statistical learning, etc.

5.1.2 Space Intelligent Decision-Making and Control Requirements

Spatial intelligent decision-making is the allocation between unlimited needs (goals and tasks) and limited resources. From the level of task, decision-making can be divided into task planning at the upper level and control at the lower level.

The main goal of autonomous mission planning for space vehicles is to independently select a series of orderly activities to form a planning scheme to complete the pending space tasks by effective reasoning based on the mission requirements of satellites, the status of equipment, the actions that satellites can take and their effects, and under the conditions of meeting time requirements and resource constraints. Spatial intelligent control, which can realize end-to-end control from decision to control, is a decision-making and control method with intelligent reasoning and solution generated by applying relevant theories and methods of artificial intelligence and integrating traditional mathematical models and methods. It is typical feature is that it can model complex decision-making and control problems by applying symbolic reasoning, qualitative reasoning and other methods under uncertain, incomplete, and fuzzy information environment reasoning and solving [5].

Intelligent decision-making focuses on the use of intelligent means to support the selection of task level and strategy level and the formulation of action sequence; intelligent control focuses more on the decision-making of dynamic systems within the task, including trajectory planning and feedback control.

The demands of satellite systems for intelligent decision-making/autonomous mission planning include autonomous selection and sequencing of detection targets by deep space detectors, organization of construction and installation sequence by on-orbit construction system, arrangement of the construction process by catalog construction, etc. The applications of satellite systems for intelligent control include intelligent motion planning, motion control, force control, intelligent autonomous navigation of detectors, autonomous rendezvous, and docking control for space robots under uncertain conditions. Compared with traditional control methods, intelligent control has better processing ability and adaptability to uncertain conditions through reasoning and learning.

The key technologies involved in spatial intelligent decision-making and intelligent control include intelligent decision support system (IDSS), strategy tree, knowledge representation, symbolic intelligence, reinforcement learning, deep reinforcement learning, Monte Carlo search method, RRT search method, adaptive dynamic planning, etc. [6, 7].

5.1.3 Space Cluster Intelligence Requirements

Clustering behavior generally exists in nature. Birds, fish and ants in the macro world, and bacteria, fungi, and some tissue cells in the micro world are all carrying out group activities. Cluster behavior has outstanding advantages in the formation of intelligence. Although individuals have limited perception ability and follow a simple movement mechanism, through the organic interaction, coordination, and control between individuals, the whole group can show complex movement behavior to form group intelligence.

Cluster intelligence refers to the intelligent characteristics of complex and orderly group behavior by many nonintelligent/simple intelligent individuals gathered in a certain space, using the self-organization cooperation mechanism characterized by stimulating work [8–10].

The theory of satellite system cluster intelligence has potential application value for formation satellite groups, formation flying robots, deep space exploration, and base construction cluster robots in the future. The emergence of swarm intelligence reduces the requirements for individual complexity, breaks the strict constraints imposed by the space environment on the individual capabilities of a single satellite, space robot or other space vehicles, and improves the group's task execution ability by means of higher adaptive ability, cooperative ability, autonomy, and other cluster behaviors.

The key technologies involved in cluster intelligence include multi-agent theory (multi-agent), cluster modeling method (particle swarm model, Vicsek model, Boid model, etc.), cluster control method, machine learning method, formation technology, multi-robot cooperation technology, etc.

5.1.4 Space Intelligent Interaction Requirements

In space, the interaction between people and people, between people and robots, and between robots and robots is limited. Intelligent interaction devices and methods can improve the control efficiency of people and robots in space environment or be conducive to the emotional adjustment of people.

Spatial intelligent interaction refers to the efficient information transmission technology between people and computers, robots, and agents by using intelligent methods for information recognition and processing through voice, expression, gesture, EEG, VR/AR, and other interactive means.

The application scenarios of space intelligent interaction include space station intelligent companion robots, robot astronauts, manned satellite voice command systems, satellite manned exploration robots, etc.

The key technologies involved in spatial intelligent interaction include natural language processing, speech recognition, gesture recognition, speech translation, EEG recognition, VR/AR technology, deep learning, etc.

5.1.5 Space Intelligent Design Requirements

Intelligent design of satellite systems refers to the technology that assists human designers to complete optimization, layout, analysis and verification by intelligent means in the process of satellite system design to achieve the purpose of improving efficiency, reliability, and design performance.

Since the design process is located on the ground, all AI technologies available for civil products have the conditions for application, so we can first break through the limitations of space conditions and apply them to the aerospace field.

The applicable aspects include structural weight reduction optimization, mechanical characteristics optimization, EMC design, antenna design, line layout optimization, controller-aided design, etc.

Key technologies involved in the intelligent design of satellite systems include genetic algorithms, simulated annealing algorithms, reinforcement learning, deep learning, etc.

5.2 Challenges and Solutions of Artificial Intelligence in Aerospace Applications

When AI's computing power, analytical power, insight, and so on surpass human beings, AI will provide better solutions in many fields than human beings. Now the design of the satellite system has initially shown "weak intelligence." The challenges and solutions of artificial intelligence technology in aerospace applications are embodied in hardware, software, and system [11, 12].

5.2.1 Hardware Level

The hardware layer requires cutting-edge hardware devices. The combination of aerospace and artificial intelligence, in addition to the design of intelligent algorithms at the software level, the most significant difference is the difference between artificial intelligence support systems. Unlike ground support systems, on-board systems have many limitations on processing performance, transmission speed, storage capacity, and energy supply.

1) *Processor*

   Taking the ground system as an example, the implementation of AI algorithms needs strong computing power support, especially the large-scale use of deep learning algorithms, which puts forward higher requirements for computing power. The deep learning model has many parameters, large amount of calculation, and larger scale of data. In the early model of speech recognition using deep learning algorithm, there are 429 neurons in the input layer, the whole network has 156 trillion parameters, and the training time exceeds 75 days; Alpha Dog has 1202 CPUs and 176 GPUs; the Google Brain project built by Andrew Ng and Jeff Dean, the leaders of artificial intelligence, uses a parallel computing platform containing 16000 CPU cores to train a deep neural network with more than 1 billion neurons. Next, if we simulate the nervous system of the human brain, we need to simulate 100 billion neurons, and the demand for computing power will increase by several orders of magnitude. Currently, the processor platforms that can be used for deep neural network construction and calculation include CPU, GPU, FPGA, and ASIC.

2) *Transmission*

The third AI climax led by the deep neural network also belongs to big data intelligence. The neural network needs to accept a large amount of sensor data every moment. Take autopilot as an example, the data generated by cameras, radars, and other devices is more than 10 GB per second. These data need to be transmitted to the designated location in time; otherwise, the network cannot provide decision support to the controller in time. At present, the bus technology used in the aerospace field is far from meeting the transmission requirements. It needs to be upgraded or reduced by some means [13–15].

3) *Storage*

A large amount of data is generated and transmitted by sensors in real time, and valuable historical data should be stored in the memory. In addition, the parameter data used by the deep neural network to record the weight value is also huge. The storage speed and storage capacity per unit volume of the ordinary memory is not satisfactory. Therefore, the use of the multi-layer neural network in satellite systems also puts forward higher requirements for memory.

4) *Energy*

Deep learning is to complete classification, regression, and other tasks through many data calculations. A more appropriate term is used to describe it as "violence aesthetics." When the existing memory is used for such violence calculations, the single energy problem is not affordable for on-board power supply and distribution. The energy consumption of AlphaGo, which defeated Li Shishi, is about 173 000 J/s. According to the daily energy consumption of 2500 kcal, the power of AlphaGo is more than 1000 times that of human beings, and human beings are the consumption of multiple physiological functions. AlphaGo is only a single Go operation.

5.2.2 Software Level

The software layer needs efficient new algorithms. Artificial intelligence software, represented by deep reinforcement learning (DRL), can be applied in the aerospace field in two steps: first, apply deep reinforcement learning to ground tests; second, apply the ground test to the space environment. At present, DRL technology has shown great potential for learning control of space robot tasks, but there are still many challenges in expansion and stability. There are still many problems that have been criticized by people, such as its convergence, interpretability, and reliability. Many researchers are exploring more efficient machine learning methods that can replace deep learning. Hinton, the father of deep learning, also made a subversive statement: "Deep learning needs to start anew and completely abandon reverse propagation." The challenges of applying DRL technology to the aerospace field are as follows [16, 17].

1) *Sample efficiency*: Although DRL algorithm provides a general framework for agents to learn high-dimensional control strategies, it usually requires millions of training samples. This makes it infeasible to directly use DRL algorithm to train agents in real scenes because it is relatively expensive to obtain empirical samples in real satellite system control. Therefore, it is very important to design an efficient sample algorithm.

2) *Strong real-time requirements*: Without special computing hardware, a very deep network with millions of parameters may be relatively slow to forward and may not meet the real-time requirements for controlling real satellite systems. A compact representation of learning agile strategies is desirable.

3) *Safety considerations*: Real satellite systems, such as Mars rover and space station manipulator, will operate in highly dynamic and potentially dangerous environments. For the wrong prediction different from the perception model, a wrong output may lead to serious accidents. Therefore, when deploying control strategies on actual autonomous systems, attention should be paid to combining the uncertainty of possible results with actual considerations.

4) *Stability, robustness, and interpretability*: DRL algorithms may be relatively unstable, and their performance may vary greatly among different configurations. To solve this problem, a deeper understanding of the learned network representation and strategy may help to detect the scenario of confrontation to prevent the satellite system from being threatened by security.

5) *Lifelong learning*: At different times and in different scenes, the environmental perception of space probe navigation or space station manipulator operation will change, which may hinder the implementation of the learned control strategy. Therefore, the ability to continue learning to adapt to environmental changes and the ability to maintain solutions to the environment that has been experienced is of key value.

6) *Generalization between tasks*: At present, most algorithms are designed for a specific task, which is not ideal because the intelligent satellite system is expected to complete a group of tasks and conduct the minimum time of total training for all considered tasks.

7) *The problem of standardizing goals*: In reinforcement learning, the expected behavior is implicitly obtained by the reward function. The goal of reinforcement learning algorithm is to maximize the accumulated long-term rewards. Although in practice, it is usually much simpler than getting the behavior itself, it is extremely difficult to define a good reward function in reinforcement

learning of satellite systems. Learners must observe the variance in the reward signal to improve the strategy. If you always receive the same return, you cannot determine which strategy is better or closer to the best.

With the continuous deepening of research on artificial intelligence theories and methods represented by DRL, human beings will achieve the goal of "solving intelligence and solving everything with intelligence" soon [18–20].

5.2.3 System Level

The system layer needs cutting-edge technological innovation. The key limitations of the combination of aerospace and intelligence include: limited resources, small data samples, unacceptable black-box status, and reliability cannot be guaranteed. In the long run, it needs innovation to completely solve the problems of on-board AI computing, transmission, storage, and energy consumption. Innovation depends on time and the development of human technology. However, the demand for intelligence in the aerospace field is imminent. In the short term, the combination of intelligence and aerospace can no longer use the directly degraded civil cutting-edge technology, but directly go deep into "the first use of cutting-edge technology achievements in aerospace," form the intelligent aerospace engineering theory, and reasonably plan the technology application mode from the top structure to avoid bottleneck.

Promote new and efficient hardware products: take the "small and complete" robot system as the pilot, actively introduce and verify new high-performance products to support the application of artificial intelligence in the satellite field, including large-scale FPGA, AI special chip, high-speed main line, full flash memory array, etc., on the premise of ensuring reliability [21].

Transfer large-scale computing to the ground: transfer the computing with low real-time requirements to the ground or to the orbiter with relatively large computing resources for execution, and form the space–space integration and space–orbit integration AI computing link. However, this will bring new problems of increasing transmission data. Therefore, it is necessary to optimize the allocation of heaven and earth resources to form a high-performance computing organism as much as possible. In addition, many optimization problems in the satellite design process can use AI technology to improve efficiency, performance, and quality. This part is completed on the ground, so the ground intelligent design technology can be given priority.

Small-scale neural networks or other intelligent machine learning methods are used for special applications: before the rise of deep learning, Curiosity Mars rover and some aircraft have adopted some on-board software with independent decision-making capabilities, such as ASE, APGEN, MAPGEN, etc. These softwares use sign intelligence, etc., and use a series of logic to address possible situations. This kind of algorithm consumes fewer computing resources and can be used in some basic judgment tasks. In addition, for data sets that are not too complex, a few levels of neural networks can meet the requirements, which is also the function that the current on-board resources may complete. The higher goal is to achieve intelligent learning detection, such as the scientific goal of OASIS to find new albedo characteristics by training itself.

Design for smart satellites/detectors: The design and implementation of smart satellites/detectors in the future have the following characteristics: definable requirements, reconfigurable hardware, reconfigurable software, reconfigurable functions, and learnable detection.

Formation of human-machine integration intelligence: with the help of human–machine integration intelligence, the intelligent behavior of the satellite is formed into a closed-loop system, which is limited to the range completely controlled by the operator. The advantages of machine computing are speed and precision, while the advantages of human thinking are flexibility and adaptability. Combining their respective advantages, human–computer integrated intelligence can be formed. By learning from each other's advantages, the demand for hardware can be reduced, and a high-level integrated intelligent system can be realized at a lower cost.

AI is a strategic technology leading the future and will be a subversive force in the field of national security. Its impact can be compared with nuclear, aerospace, information, and biotechnology. The space power led by NASA of the United States has taken space as an important stage for AI to play its role. The need for intelligence in the aerospace field is imminent: without intelligent autonomy in the field of deep space exploration, it will be impossible to move forward in the future; manned satellites begin to develop and configure intelligent robot assistants in sealed cabins to assist astronauts' on-orbit activities; the in-orbit service satellite will develop a noncooperative satellite capture system with computer vision and cognitive reasoning as the core; intelligent autonomous operation or supervised autonomous operation will become an important development direction of the future space system; the development of intelligent military aerospace integrates many cutting-edge intelligent technologies, and so on [22].

In a word, in order to seize the commanding heights of future space, we will deeply explore the integration of artificial intelligence and aerospace technology, take the strength of all people, and vigorously develop the software, hardware, and system cutting-edge technologies such as space intelligent perception, space intelligent decision-making and control, space cluster intelligence, space intelligent interaction, space intelligent design, and so on, and support the future of intelligent space.

References

1 JACEK KRYWKO (2018). *To Make Curiosity (et al.) More Curious, NASA and ESA Smarten Up AI in Space*. Ars Technica.

2 Vassev, E. and Hinchey, M. (2014). On the autonomy requirements for space missions. In: *16th IEEE International Symposium on Object/component/service-oriented Real-time distributed Computing*. Paderborn, Germany: IEEE.

3 McCarthy, J. (2006). A proposal for the Dartmouth summer research project on artificial intelligence. *AI Magazine*. 27 (4): 12.

4 Cichy, B. (2010). *Spacecraft Flight Software (FSW-10), at Workshop*. Goddard Space Flight Center.

5 Tony Greicius, A.I. (2017). *Will Prepare Robots for the Unknown*. Pasadena, California: *Jet Propulsion Laboratory*.

6 Ren, S., Girshick, R., Girshick, R. et al. (2017). Faster R-CNN: to-wards real-time object detection with region proposal networks. *IEEE Transactions on Pattern Analysis & Machine Intelligence* 39 (6): 1137–1149.

7 Cornelius, R. (2016). International Space Station (ISS) payload autonomous operations past, present and future, *14th International Conference on Space Operations*, Daejeon, Korea, (16–20 May 2016). SpaceOps 2016, p. 2594.

8 Wong, C., Yang, E., Yan, X.T., and Gu, D. (2017). Adaptive and intelligent navigation of autonomous planetary rovers – a survey. In: *NASA/ESA Conference on Adaptive Hardware and Systems (AHS)*, 237–244. IEEE.

9 Liewer, P.C., Klesh, A.T., Lo, M.W. et al. (2014). A Fractionated space weather base at L5 using CubeSats and solar sails. In: *Advances in Solar Sailing* (ed. M. Macdonald), 269–288. Berlin Heidelberg: Springer-Verlag.

10 Hall, L. (2015). *Edison Demonstration of Smallsat Networks (EDSN)*. NASA.

11 Liewer, C., Klesh, T., Lo, W. et al. (2014). A fractionated space weather base at L5 using CubeSats and solar sails. In: *Advances in Solar Sailing* (ed. M. Macdonald), 269–288. Heidelberg: Springer Berlin.

12 Baird, D. (2017). NASA explores artificial intelligence for space communications. www.nasa.gov/scan (accessed 9 December 2017).

13 Chongjie, Z. (2016). Co-optimization multi-agent placement with task assignment and scheduling. In: *Proceedings of the Twenty-Fifth International Joint Conference on Artificial Intelligence (IJCAI)*, 3308–3314.

14 Mainprice, J., Hayne, R., and Berenson, D. (2016). Goal set inverse optimal control and iterative replanning for predicting human reaching motions in shared workspaces. *IEEE Transactions on Robotics* 32 (4): 897–908.

15 Ewerton, M., Neumann, G., Lioutikov, R. et al. (2015). Learning multiple collaborative tasks with a mixture of interaction primitives. In: *IEEE International Conference on Robotics and Automation (ICRA)*, 1535–1542. IEEE.

16 Jiang, Y. and Saxena, A. (2014). Modeling high-dimensional humans for activity anticipation using gaussian process latent CRFS. In: *Robotics: Science and Systems.*

17 Huang, C.-M. and Mutlu, B. (2016). Anticipatory robot control for efficient human-robot collaboration. In: *ACM/IEEE International Conference on Human-Robot Interaction (HRI)*, 83–90. IEEE.

18 Koppula, H.S., Jain, A., and Saxena, A. (2016). "Anticipatory planning for human-robot teams," *Experimental Robotics: The 14th International Symposium on Experimental Robotics*, vol. 109, no. i, pp. 453–470. Springer International Publishing.

19 Knepper, R.A., Mavrogiannis, C.I., Proft, J., and Liang, C. (2017). Implicit communication in a joint action. In: *12th ACM International Conference on Human-Robot Interaction.* Vienna, Austria: IEEE.

20 Ghadirzadeh, A., Bütepage, J., Maki, A. et al. (2016). A sensorimotor reinforcement learning framework for physical human-robot interaction. In: *2016 IEEE/RSJ International Conference on Intelligent Robots and Systems (IROS)*, 2682–2688. IEEE.

21 Mnih, V., Kavukcuoglu, K., Silver, D. et al. (2015). Human-level control through deep reinforcement learning. *Nature* 518 (7540): 529–533.

22 Zhang, M., McCarthy, Z., Finn, C. et al. (2016). Learning deep neural network policies with continuous memory states. In: *2016 IEEE International Conference on Robotics and Automation (ICRA)*, 520–527. IEEE.

6

Intelligent Remote-Sensing Satellite System

6.1 Technical Analysis of Intelligent Remote-Sensing Satellite System

6.1.1 Technical Requirements Analysis of Remote-Sensing Satellite Artificial Intelligence System

As space remote-sensing technology plays an increasingly important role in national economic construction, its deficiencies in technology and operation and control mode are increasingly prominent, which restricts the overall effectiveness of satellites. Mainly reflected in the following aspects:

First, the increasing ability and volume of remote-sensing satellite data acquisition have brought great pressure on satellite downlink data transmission and post-processing. The continuous improvement of remote-sensing satellite imaging performance, such as spatial resolution, spectral resolution, and scanning width, also increases the amount of satellite data in geometric progression. For example, NASA receives more than 3.5 TB of observation data every day, and only a small part of it is used [1].

Second, the ground environment background and observation targets are complex and diverse. Even though the remote sensing characteristics of the same type of ground objects change with the seasons, the indicators of satellite payload development and launch are basically unchanged, such as Landsat/TM, SPOT, MODIS, EO-1, and other satellites.

Thirdly, according to the analysis of the development of deep space exploration, remote sensing is the main method of deep space exploration. Remote-sensing image is the most direct, effective and important means to obtain the information of deep space objects [2–4]. Due to the particularity of deep space environment,

Intelligent Satellite Design and Implementation, First Edition. Jianjun Zhang and Jing Li.
© 2024 The Institute of Electrical and Electronics Engineers, Inc. Published 2024 by John Wiley & Sons, Inc.

special requirements are put forward for deep space exploration and remote-sensing activities, mainly including the following aspects:

1) The characteristics of the remote sensing detection object are unknown, and appropriate detection mode needs to be adopted.

   The purpose of deep space exploration is to detect the unknown information of extraterrestrial objects. Therefore, unlike earth observation, it lacks a lot of prior knowledge, which may lead to inefficient or even ineffective remote sensing detection modes. A detection system that combines multiple modes and can be switched automatically in orbit is required. For example, for some large flat areas, it is of little significance to use high-resolution spectral detection and high-density laser lattice detection. Or some areas need infrared detection, which is not suitable for visible light detection, etc. Therefore, it is necessary to evaluate the telemetry detection mode in orbit and switch it in time to improve detection efficiency [5, 6].

2) The deep space exploration environment is complex and changeable, and reasonable parameters need to be adopted.

   Due to the distance between the deep space detector and the earth, there are great difficulties in high-precision measurement and control on the ground. Therefore, it is impossible to carry out high-precision prediction and take effective human intervention measures for the orbit conditions, lighting environment, electromagnetic environment, etc., of the detector, and the deep space detector needs to face complex and changeable space environment. Currently, it is necessary to have certain adjustable parameters and automatic parameter adjustment ability to use the detection requirements at that time. For example, for areas with low illumination conditions, the visible light detection gain needs to be adjusted accordingly to obtain high-quality images. The detection resolution requirements of key areas and non-key areas are different, and the remote sensor needs to have a certain zoom ability to adapt to different resolution requirements [7, 8].

3) Remote sensing platform resources are limited, and detectors need to be used reasonably and efficiently.

   Due to the long distance and limited carrying capacity of deep space exploration, the detector platform is generally small and cannot provide a large amount of resources, including mass, size, and power consumption. Therefore, the intelligent remote sensor for deep space exploration should reasonably allocate platform resources according to priority. For example, for some active detection loads, such as laser, microwave, neutron, etc., due to high energy consumption, the platform may not be able to meet its long-term operation. Therefore, from the perspective of detection needs and data availability, it is necessary to independently evaluate the remote sensing efficiency, reasonably select detectors and working hours, and optimize the use of platform resources.

4) Remote-sensing data transmission capacity is limited, and on-orbit data processing is required.

   To obtain more accurate and comprehensive target information, the detector will produce a huge amount of data. However, the data transmission capacity is limited, resulting in a large amount of data that cannot be downloaded. Therefore, the remote sensor is required to have a strong ability to automatically process in-orbit data, automatically screen useful data, and reduce the amount of data transmitted. In order to solve a variety of complex problems that may be encountered in deep space exploration and remote-sensing activities, improve the ability to acquire deep space remote-sensing data and the effectiveness of remote-sensing data, and lay a technical foundation for the efficient development of deep space exploration activities in the future, it is necessary to carry out intelligent remote sensing research on deep space targets in three aspects: intelligent acquisition mode of deep space exploration information, intelligent adjustment of deep space exploration remote sensor parameters and intelligent processing of deep space exploration on-orbit data [9, 10].

Fourth, the satellite's mission implementation process is mainly based on the telemetry and telecontrol system with large time delay between the space and the earth. First, the mission is planned on the ground, and the instructions are annotated. The satellite carries out the relevant tasks in strict accordance with the instructions at the specified time. The mission implementation effect and onboard status are transmitted to the ground through telemetry data, which is interpreted by professionals and based on this, the new mission is planned. This planning method has certain disadvantages. For example, it takes a long time from the start of the planning process to the generation of a new planning scheme. If the mission plan is wrong in the implementation process, the corresponding time of the new plan may take several hours. Similarly, the delay of the mission may also make the satellite miss a better observation opportunity. In addition, there will also be some errors in the estimation of the system status, which may lead to mission failure. Compared with the autonomous closed-loop decision-making, planning, and implementation on the satellite, the satellite's emergency response ability, observation income and resource utilization efficiency will be seriously restricted under the current limited measurement and control capability [11, 12].

Finally, the automation of data processing is low, and the timeliness of information extraction is poor. The traditional model of "satellite data acquisition – ground station receiving and processing – data distribution – professional application" cannot meet the timeliness requirements of future high mobility, high response to acute disaster emergency monitoring and some military applications. It is necessary to quickly convert data into information and send it to end users at different levels in real time or quasi-real time. At present, the ability of remote-sensing satellite data acquisition has been continuously enhanced, with spatial resolution ranging from kilometers to tens of meters or even submeters. The spectral resolution has been improved from hundreds of nanometers to nanoscale (such as HYPERION, Compact High-Resolution Imaging Spectrometer (CHRIS)). Relative to the increasing satellite data acquisition capability, the level of automatic data processing is relatively low: from data acquisition to information extraction and then to the application of products, there are many manual interventions. Some remote-sensing applications such as information extraction are mainly based on artificial visual interpretation, and information services are difficult to guarantee timeliness. In fact, people's demand for satellite products has shifted from the image itself to more and more diversified thematic information. Ordinary users do not pay attention to the complex technology and details of image processing. Satellite real-time processing and downloading of professional application products has become one of the core concepts of future satellite design.

The traditional satellite remote sensing technology has been difficult to meet the needs of fast, accurate, and flexible remote sensing data acquisition and information product production. It is necessary to build an intelligent remote-sensing satellite system with the ability to work imaging mode optimization, rapid information production, and transmission. Taking the design of remote-sensing satellite systems and the development of ground information processing technology into consideration is an important frontier direction for the development of earth observation satellite technology with high spatial, hyperspectral, and high radiation resolution [13–15].

6.1.2 Concept Connotation of Intelligent Remote-Sensing Satellite System

Intelligent remote sensing is an interdisciplinary subject of remote sensing technology and intelligent science. The intelligence of the remote sensor is mainly reflected in its ability to judge, learn, and update itself. It is an intelligent expert system, which can switch the detection mode, switch the parameters of the remote sensor, and intelligently process and transmit the remote sensing data according to the analysis results of the data to realize the autonomous detection of unknown targets. In the complex and changeable space environment, the usefulness and efficiency of remote-sensing data can be improved by optimizing the resource ratio of each system independently.

Since the 1960s, space remote sensing has become one of the most important means for human observation of the earth. After decades of development, intelligent-sensing satellites have gradually developed into multi-platform, multi-sensor, multi-angle, and high-resolution features. With the development of information technology, people's requirements for remote-sensing have also developed from image-based data to image-based information and knowledge. Developing the next generation of intelligent remote-sensing satellites has become an urgent task.

Intelligent remote-sensing satellite system, through satellite autonomous mission planning, remote sensing data on-orbit processing, and multi-load collaborative control technology, enables satellites to realize multi-satellite mutual guidance and multi-load collaborative work for specific tasks, realize multi-type satellite system collaborative earth observation, multi-source information fusion processing on satellite, direct distribution of information products, and improve satellite application capabilities.

The intelligent remote-sensing satellite system can not only effectively solve the problems existing in the current remote-sensing satellite and its application and meet the needs of all customers but also has the advantages of real-time adjustment of imaging control strategy, data processing and analysis, high-precision attitude determination and positioning capabilities and image processing capabilities. Therefore, it is of great significance to carry out research on intelligent remote-sensing satellite system, which can provide reference for the design and development of remote-sensing satellite in the future [9, 16].

6.1.3   Main Features of Intelligent Remote-Sensing Satellite System

In recent years, some major aerospace remote sensing countries have attached great importance to the development of real-time processing technology of onboard data and have begun to develop some satellites with some intelligent features, such as the Naval EarthMap Observer (NEMO) satellite of the United States Navy, the BIRD satellite of the German Aerospace Agency, the Co-ordinated Constellation of User Defined Satellites (COCONUDS) system of Europe, and the PROBA satellite of the European Space Agency. Compared with traditional earth observation satellites, these satellite systems generally have the following characteristics.

6.1.3.1   Special Function
Intelligent satellites are not designed for the main purpose of data acquisition but for some professional applications and special needs. The hyperspectral imager and 5 m resolution panchromatic camera carried by the US NEMO satellite is mainly used for rapid analysis of coastal zone environment, extraction of surface reflectance, sea surface remote sensing reflectance, water turbidity, chlorophyll, CDOM, suspended solids, water depth, underwater reflectance, etc. The FOCUS platform on the International Space Station is used for automatic fire detection. FOCUS comprehensively uses forward-looking camera, high spatial resolution

camera, and Fourier infrared imaging spectrometer to realize potential fire detection, solar flare or cloud removal, warm surface removal, hot spot clustering, and other functions. The BIRD satellite of the German Space Agency is mainly used for the detection and evaluation of thermal anomalies such as forest fires, volcanic eruptions, oil field, and coal mine fires. In addition, automatic monitoring of flood disaster is also one of the directions of intelligent satellite design in the future.

6.1.3.2 Variable Imaging Mode

The CHRIS on the PROBA small satellite launched by ESA can obtain hyperspectral images from five angles ($-55°$, $-36°$, $0°$, $36°$, and $55°$) within 2.5 minutes and can realize the conversion of five different spatial and spectral resolution imaging modes according to different observation objectives and application requirements such as water, vegetation, and land. It can obtain 62 bands and 34 m spatial resolution images in the 411–997 nm spectral band and can also change the imaging mode to obtain 18 bands and 17 m spatial resolution hyperspectral images in the 411–1019 nm spectral band according to the needs of water quality remote sensing.

6.1.3.3 Real-time Processing of Onboard Data

Real-time processing of onboard data is one of the most prominent features of intelligent satellites. The amount of processed information product data is greatly reduced, which can reduce the pressure of decimal transmission and make remote sensing information directly received by end users. The hyperspectral data processing of NEMO satellite in the United States uses the adaptive spectral recognition system ORASIS, and its data processing uses convex set analysis and orthogonal projection transformation technology to decompose and generate 10–20 end elements for specific scenes to realize automatic data analysis, feature extraction, and data compression. The in-orbit data processing capability of BIRD satellite includes radiometric correction, geometric correction, disaster early warning, and monitoring thematic information production, etc.

6.1.3.4 Real-time Data Download

The European COCONUDS constellation is composed of 10 polar orbiting satellites, which are used for environmental monitoring and do not store data on the satellite to realize real-time data download. France's Pleiades satellite data is compressed and transmitted down at a speed of up to 4.5 Gbit/s.

In addition to the above intelligent remote-sensing satellites with onboard real-time processing performance, some commercial satellites that have been successfully launched or are in the future planning also have more intelligent features. For example, Digital Globe launched the WorldView-1 panchromatic image satellite with a resolution of 0.5 m in September 2007, which has excellent rapid response capability and can conduct up to 24 continuous multi-angle observations of the same target according to the demand. The WorldView-2 launched subsequently has eight multi-spectral bands and one panchromatic band. With its powerful attitude control gyros (CMGs), it can flexibly realize wide coverage (multiple back and forth scanning of 65.6 km), long strip (16.4 km) or stereo observation (48 km wide) imaging modes according to the needs, greatly improving the efficiency of high-resolution satellite data acquisition. SPOT 6 and SPOT 7 satellites launched by SPOT also have greater flexibility in data acquisition. Combined with long-term and short-term weather forecasts, they reduce the acquisition of invalid data in rainy weather conditions and optimize the data acquisition process [9, 17].

6.2 Basic Structure of Intelligent Remote-Sensing Satellite System

"A satellite remote-sensing load system scheme based on artificial intelligence technology" combines artificial intelligence technology to design the overall scheme of intelligent satellite remote-sensing load and the scheme of intelligent expert system and intelligent execution system, and takes river basin observation as an example to introduce the working process of intelligent satellite remote-sensing load system.

The intelligent remote-sensing satellite payload includes three parts: intelligent expert system, intelligent execution system, and intelligent semantic interpretation system, as shown in Figure 6.1. The intelligent expert system is composed of

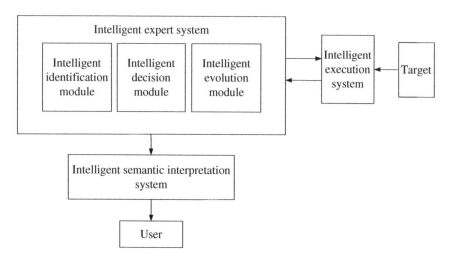

Figure 6.1 Structural composition of intelligent remote-sensing satellite load system.

intelligent identification module, intelligent decision-making module, and intelligent evolution module. Intelligent execution system mainly refers to new remote sensors, including full-field sensing instrument, zoom medium/high-resolution imager and variable spectral resolution imager. The intelligent semantic interpretation system is mainly used for the barrier-free communication of users [8, 18, 19].

The intelligent expert system is the brain of the remote sensor system. Through the intelligent recognition module, it carries out in-orbit analysis on the data obtained from the predetection of the full-field sensing instrument, compares the geometric features in the image with the features in the feature database, extracts the target features and identifies them, and judges the target attributes. The intelligent decision module determines the value of the target and gives the best imaging parameters of the remote sensor, which is used to command the intelligent execution system. The intelligent execution system receives the instructions of the intelligent expert system, drives the zoom medium/high-resolution imager and the variable spectral resolution imager to track and detect the specified target. At the same time, the work of the intelligent expert system needs to carry out self-evaluation and ground up evaluation. Through the analysis of the historical evaluation data, the experience of "numerical" is summarized to guide the next expert system work, that is, the intelligent evolution module [20–22].

6.2.1 Intelligent Expert System

6.2.1.1 Intelligent Identification Module

The function of the intelligent recognition module is to automatically analyze the image information by means of computer information processing, so as to find the object of interest and confirm the object type.

The intelligent recognition module is divided into three parts: image target area capture, image feature extraction, and image feature recognition. The target area capture link extracts the valuable areas in the remote-sensing image from a large number of useless information. The image feature extraction link extracts the target features from the valuable target area information. The image recognition module classifies and describes the image according to the geometric and texture characteristics of the figure using recognition theories such as pattern matching and discriminant function [22–24].

6.2.1.2 Intelligent Decision Module

What kind of working mode should the remote sensing system adopt for a certain target, and whether the parameter setting is the best, are the system problems of the optimal target value evaluation and stress plan planning.

In a remote sensing image, because there may be multiple targets at the same time, some of them are more concerned by users, such as airports and ports, while others can be ignored. The importance of these objectives can be quantitatively expressed by the objective preference function. First, the output of the target capture phase, namely the recognition vector, is input into the target preference function, respectively, and the system determines which targets can be ignored and which need to be tracked. Then the system extracts the important objects and retrieves the cases in the case database to detect the similarity. According to the modification rules, the historical decision is modified with reference to the current environment, and then the quasi-decision vector is obtained.

The decision indicators are extracted from the pseudo-decision vector, and the corresponding detection mode, direction, and spectral segment combination planning scheme is estimated by a specific algorithm to determine whether it meets the target threshold. If it does not meet the target threshold, the above process is repeated, and the mode planning is repeated until the target threshold is met, and the next step of the parameter and path stress planning system is entered.

The parameter and path stress planning system include the establishment of a change algorithm, obtaining the function based on the parameter and path preference vector, and introducing the stress imaging simulation vector based on the parameter and path, and obtaining the optimal parameter and path combination through the specific algorithm cycle parameter and path planning. The best mode, parameter and path planning scheme are used as the final onboard stress decision vector, enter the intelligent execution system, and return to the correction case database.

6.2.1.3 Intelligent Evolution Module

One of the main characteristics of intelligent systems is that they can adapt to unknown environments. Learning ability is one of the key technologies of intelligent systems. The intelligent evolution system on the satellite needs to have the ability of self-judgment, self-learning, and self-renewal. Among them, self-learning and self-renewal are the feedback based on image information obtained by the intelligent system after deciding on the satellite and taking a stress response to independently judge the effect of a certain decision and stress response. If it reaches the default threshold of the system, that is, the decision and response are correct, then the decision and stress response will be updated into the strategy database for the next visit. This intelligent evolution scheme is based on reinforcement learning technology, and is a special learning method that takes environmental feedback as input and adapts to the environment. For reinforcement learning, its goal is to learn a stress strategy in the new environment.

6.2.1.4 Intelligent Execution System

The intelligent execution system receives and executes the instructions of the intelligent expert system to complete the imaging task. The intelligent execution system includes large-area low-resolution imaging, local medium-resolution imaging, and high-resolution imaging of key targets. Among them, low-resolution imaging realizes wide area perception and automatically locates the target area. Medium-resolution imaging realizes the medium-resolution search of the target area and the positioning of key targets. High-resolution imaging realizes high spectral resolution recognition of key targets.

6.2.2 Intelligent Semantic Interpretation System

Intelligent semantic interpretation system is to complete the process of image digitization through information evaluation, image fusion, and data compression in orbit and directly distribute the simplified data information to the user's hand-held device and command center in a "point-to-point" manner through onboard communication means to provide users with visual real-time data information and improve their ability to perceive emergencies.

In image semantic interpretation, data characteristics is analyzed first, then features are extracted, and the relationship between features and semantics is built through one-to-one mapping relationship to generate semantic content and features. Finally, learning and training are conducted according to the relationship between semantic content and features, and a knowledge-based semantic model is established, and finally the image interpretation results using semantic representation are obtained. This method can effectively improve the association relationship between different features, provide more association concepts for the mining of potential knowledge, make the understanding of target objects more comprehensive and accurate, and achieve effective information mining and intelligent interpretation of massive data [25, 26].

6.3 Key Technical Directions of Intelligent Remote-Sensing Satellite System

Compared with traditional imaging methods, the imaging method of satellite remote-sensing payload system based on artificial intelligence technology has the following advantages: (i) imaging on demand, more targeted; (ii) low-resolution perception ability, high-resolution key target detailed investigation ability; (iii) reduces the amount of downlink data and reduce the pressure of satellite information transmission; (iv) high timeliness of information utilization; (v) it has the function of learning and evolution [27].

The satellite remote sensing payload system based on artificial intelligence technology can systematically design the whole link of intelligence collection, processing, distribution, and application from the source of data acquisition, improve the timeliness and usefulness of information, and provide effective support for users to obtain information on demand and in time. It can be applied to large-scale remote-sensing imaging, automatic search, identification and positioning of interested targets, tracking and detailed investigation of dynamic and static targets, and has independent situation awareness and intelligence processing capabilities. It has great application value in the fields of real-time response on orbit, rapid positioning and interpretation of emergencies, and rapid capture and tracking of moving targets.

6.3.1 Develop Autonomous Parameters and Mode Adaptation Technology of Remote Sensor

The future intelligent remote-sensing payload needs to change the spatial resolution, spectral segment, azimuth direction, and exposure time for different targets and environments to obtain the best imaging quality. This requires efforts to develop the autonomous parameters and mode adaptation technology of the remote sensor, so that the remote sensor can adjust the imaging mode and imaging parameters according to the actual situation at that time, and improve the imaging pertinence and working efficiency of the intelligent remote sensing load.

6.3.2 Develop Satellite Autonomous Mission Planning Technology

Satellite autonomous mission planning technology can make full use of satellite resources. It is necessary to plan when, what mode and what targets to implement observation, and effectively improve the pertinence and timeliness of observation tasks. It is a key link in the development of intelligent remote-sensing payload system technology. At present, the research on this technology is still at the theoretical stage, and there is no successful application experience. Reasonable arrangement and full utilization of satellite resources to maximize user demand has become an urgent problem to be solved in the development of intelligent remote-sensing payload system technology [17].

6.3.3 Develop In-Orbit Semantic Interpretation Technology of Remote-Sensing Images

The on-orbit interpretation technology of remote-sensing images completes the transformation from data volume to information volume from the source of information acquisition, ensures the effectiveness of downlink data, improves the ability to perceive emergencies, and provides effective space-based information support for the command center. This is one of the important technical bases for gradually realizing the research on intelligent remote-sensing payload system.

In order to realize the intelligent design of remote-sensing payload system as soon as possible, it is urgent to break through the technology of semantic interpretation of remote-sensing images in orbit [28].

6.4 Typical Application Cases

6.4.1 Rapid Intelligence Generation and Release of Remote-Sensing Satellite

1) During the emergency rescue, remote-sensing satellites are required to be able to quickly intelligently process the information of the original image obtained, extract the target information and generate battlefield intelligence, and then quickly transmit the intelligence to the end user through the inter-satellite link or satellite-ground link.

2) The original image obtained by visible/infrared/SAR remote-sensing satellite is used for radiation correction, geometric correction, and specific target recognition in orbit to generate intelligence information and quickly release it to end users. Among them, radiation correction, and geometric correction can be solved by using conventional image processing algorithms. Specific target recognition can adopt artificial intelligence algorithm based on deep learning. Through large sample training, the onboard algorithm can master the ability of feature recognition and extraction for specific targets. The capacity of the original training set can be expanded through the sample expansion technology, and the recognition accuracy and recall rate of the artificial intelligence algorithm can be improved by means of the generative confrontation network and other methods [29].

3) The application of artificial intelligence algorithm can effectively improve the adaptability and accuracy of the on-orbit target recognition and tracking algorithm, increase the ability of intelligent information extraction on the satellite, shorten the time delay of information acquisition in wartime, and win the lead for me. For example, if the satellite payload image at a certain time contains a high-value emergency target, then the subsequent payload image will also have a high probability of including the emergency target. The emergency target can be identified by combining these images. Or, based on a small number of samples, the accuracy of environmental situation awareness and assessment can be improved by generating new samples, and through confrontation learning in the process of generation and discrimination. For example, the enemy forces have gathered in a certain area in a directional way. It is difficult for the traditional algorithm to perceive this emergency trend. The artificial intelligence algorithm can solve this problem.

6.4.2   On-track Feature-level Fusion Processing and New Feature Learning

1) In emergency or emergency tasks, it is often necessary to observe and analyze the key targets as carefully as possible at the fastest possible speed. The fusion processing of remote-sensing data of different types and scales obtained from multiple types of remote-sensing satellites in orbit will help improve the precision of target recognition and analysis. In addition, reconnaissance satellites should not only recognize and monitor known targets, but also can pay attention to and alarm the undefined targets with potential dangers, that is, the self-learning ability of new features of reconnaissance targets.

2) Various types of remote-sensing satellites acquire multiple types of remote-sensing satellite images for local areas and then conduct feature-level fusion processing and application of multi-source images by a main satellite or ground data center in orbit to generate high-value intelligence information. The target self-learning method based on a radial basis function network can be adopted, which has the characteristics of simple structure, strong memory ability, and fast update speed for the known target database. Through feature fusion, new features changed from defined targets can be quickly input into the database. Based on this self-learning method, the satellite has the advantages of fast learning speed and strong sensitivity in the face of unknown potential targets or new features of targets, which will effectively improve the adaptability and robustness of the satellite to target perception, help to find more potential targets, and improve the adaptability to target appearance changes and other external interference [30].

3) According to the mission data (including ground injection, inter-satellite distribution, onboard storage, etc.) and the collected multi-source payload data (visible light, infrared, SAR, etc.), the emergency target in the payload information is extracted through the target recognition and tracking module based on multi-source information fusion. At the same time, for the emerging potential emergency target characteristics, the judgment ability is improved through target characteristics and potential target self-learning module. For the environmental situation information and satellite basic capability information, the results of the environmental situation assessment of the battlefield area and the satellite's own state assessment are obtained through the environmental situation awareness and assessment module and the satellite's own state awareness and assessment module. Based on the above multi-source information and processing results, the satellite response strategy multi-source information self-learning module is used to obtain the optimal strategy of satellite response or action for the current battlefield environment situation.

6.4.3   On-track Independent Disaster Identification and Alarm

1) Restricted by the processing capacity and data transmission rate on the satellite, the traditional way of acquiring data on the satellite, the ground processing, and distribution process of taking images from the satellite need to be delayed by hours, so that users can get useful information, which cannot meet the timeliness requirements of disaster reduction and relief.

2) Based on remote-sensing image samples and using deep learning technology, a discriminant model is established for volcanic eruptions, fire emergencies, postdisaster vegetation changes, water pollution, urban building violations, and other abnormal events to realize rapid detection and alarm of abnormal events. Aiming at the problem of small samples in the identification of sudden disasters, the adversarial network is used to generate new samples. At the same time, the migration learning technology is used to reduce the sample size requirements, improve the recognition rate, and ensure the application efficiency [31].

3) It is conducive to the establishment of a monitoring system for high-temperature targets and high-risk targets, the improvement of environmental monitoring and comprehensive disaster reduction capabilities, and the reduction of national losses. Improve information timeliness, solve data transmission bottleneck, and improve load utilization.

6.4.4   Distributed Autonomous Mission Planning for Constellation

1) The constellation/constellation of remote-sensing satellites is increasingly large, and the working mode of single satellite is increasingly complex and flexible. Traditionally, the business model that the ground plans and arranges instructions for a single satellite and controls the satellite to work through instructions or command groups has come to an end in terms of cost and efficiency. Through the application of artificial intelligence means, a single satellite can have a stronger and more independent task planning ability, and through the application of distributed collaborative task planning algorithms and patterns, a multi-task unified optimization allocation and scheduling business model for star clusters can be built [32].

2) The remote-sensing satellite constellation/constellation uses a distributed method to independently distribute and execute the observation tasks issued on the ground, reduce the pressure of ground staff on the control of satellites, and realize the full use of satellite observation resources. After receiving various tasks, the satellite can independently complete the planning work such as the division of regional targets, the arrangement of multi-target timing, the planning of task execution/switching process, and the trade-off of task conflicts, in

combination with the environmental information and its own status information. In this process, the key is to solve how to quickly complete the mapping from task space to execution space, and quickly find the optimal solution in the infinite possible solution space, so as to achieve the optimal use of time. AI algorithm will probably play an important role in this optimization process.

3) The development of autonomous mission planning technology, on the one hand, can significantly improve the ability to execute complex tasks and quickly respond to emergencies; on the other hand, it can effectively break through the bottleneck of measurement and control capability and reduce the complexity of ground control. For example, through the onboard real-time information processing module, the target value can be determined and used as the input information for autonomous mission planning, so that the satellite can actively discover and detect the target area and changing targets and independently process and distribute real-time tactical information, which is used to support tactical applications that require high timeliness of information [22].

6.4.5 AI Technology-assisted OODA Loop

1) "Observation – judgment – decision – action" (OODA loop) is the core of the emergency process of intelligence collection and decision-making to achieve the commander's intention. Taking the typical mission process of space-based weapons and equipment as an example, OODA can describe the operational process including reconnaissance and detection, identification and confirmation, threat judgment and decision, target attack, and damage assessment. It is difficult to guarantee the tasks of massive data mining and multi-information association only by relying on manual interpretation and analysis. It is urgent to improve the efficiency of each link through artificial intelligence technology.

2) In the detection, identification and confirmation, threat judgment and decision-making links, the accuracy and speed of each link can be improved by using the target signal authenticity recognition based on deep learning, detection alarm, weak signal target detection, signal correlation and deblurring in the dense echo link, and the overall situation can be formed. In the decision-making (strike and evaluation) link, the results of intelligent information processing are usually used as the basis for auxiliary decision-making and are integrated with human command strategy and intelligence to give full play to the advantages of human–computer hybrid intelligence and make better decisions [33].

3) Improve the ability to quickly, comprehensively, and accurately perceive the battlefield situation. Machine intelligence is integrated with human command strategy and intelligence to optimize battlefield decision-making and improve the ability to win.

References

1 Tomar, G.S. and Verma, S. (2007). Gain & size considerations of satellite subscriber terminals on earth in future intelligent satellites. In: *2007 IEEE International Conference on Portable Information Devices*, 1–4. IEEE.

2 Ramapriyan, H., McConaughy, G., Morse, S., and Isaac, D. (2004). Intelligent systems technologies to assist in utilization of earth observation data. In: *Earth Observing Systems IX, Proceedings of SPIE*, 3922–3925. Bellingham: SPIE.

3 Chien, S., Wichman, S., Engelhart, B. et al. (2002). Onboard autonomy software on the three corner sat mission. In: *Proceedings of the International Symposium on Artificial Intelligence Robotics & Automation for*.

4 Wilson, L.T. and Davis, C.O. (1998). Hyper spectral remote sensing technology (HRST) program and the Naval EarthMap Observer (NEMO) satellite. In: *Infrared Spaceborne Remote Sensing VI Meeting*, 2–10. San Diego: SPIE.

5 Davis, C.O. (2000). Spaceborne hyperspectral technology: the Naval EarthMap Observer (NEMO) [J/OL]. http://www.scs.gmu.edu/~rgomez/EOS%20Lectures/5Lecture%2029%20Sep%2003/DoD/NEMO.pdf (accessed 31 January 2011).

6 Zhou, G. (2001). *Architecture of Future Intelligent Earth Observing Satellites (FIEOS) in 2010 and Beyond*. Norfolk: National Aeronautics and Space Administration, Institute of Advanced Concepts (NASA-NIAC).

7 Chien, S., Cichy, B., Davies, A. et al. (2005). An autonomous earth-observing sensorweb. *IEEE Intelligent Systems* 20 (3): 16–24.

8 Wilson, T.M. and Davis, C.O. (1999). Naval EarthMap Observer (NEMO) satellite. In: *Imaging spectrometry V, Proceedings of SPIE*, 2–11. Bellingham: The International Society for Optical Engineering.

9 Winfried, H. (2001). Thematic data processing on board the satellite BIRD. *Remote Sensing* 4540: 412–4191.

10 Yuhaniz, S.T., Vladimirova, T., and Gleason, S. (2007). An intelligent decision-making system for flood monitoring from space. In: *ECSIS Symposium on Bio-inspired, Learning, and Intelligent Systems for Security*, 65–71. Los Alamitos: IEEE Computer Society.

11 Barbara, V.M. and Ruddick, K. (2004). The Compact High Resolution Imaging Spectrometer (CHRIS): the future of hyperspectral satellite sensors. Imagery of Oostende Coastal and Inland Waters. In: *The Airborne Imaging Spectroscopy Workshop*. Bruges.

12 AGI (2002). *GREAS for STK User's Manual and Tutorial*.

13 Analytical Graphics Inc (2002). Satellite Tool Kit (STK). www.agi.com.

14 Chien, S., Rabideau, G., Willis, J., and Mann, T. (1999). Automating planning and scheduling of shuttle payload operations. *Artificial Intelligence Journal* 114: 239–255.

15 Myers, W.A., Smith, R.D., Stuart, J.L. et al. (1998). NEMO satellite sensor imaging payload. *Proceeding of SPIE* 3437: 29–401.

16 Davis, C.O. (1997). The Hyperspectral Remote Sensing Technology (HRST) program. In: *Proceedings of the ASPRS Meeting: Land Satellite Information in the Next Decade. II. Source and Applications*, Washington, DC, 2–5. AIAA.

17 Arnaud, M., Boissin, B., Perret, L. et al. (2006). The Pleiades optical high resolution program. In: *Proceeding of the 57th IAC/IAF/IAA, 2–6 October 2006*. Valencia: International Astronautical Congress: IAC-06-B1.1.04.

18 Verduijn, F.F., Algra, T., Close, G.J. et al. (2001). COCONUDS? Two years on. *Acta Astronautica* 52 (9-12): 829–832.

19 Davis, C.O., Horan, D.M., and Corson, M.R. (2000). On-orbit calibration of the Naval EarthMap Observer (NEMO) Coastal Ocean Imaging Spectrometer (COIS). In: *Imaging spectrometry VI, Proceedings of SPIE*, 250–259. Bellingham: The International Society for Optical Engineering.

20 Choi, S., Scrofano, R., Prasanna, K.V., and Jang, J.W. (2003). Energy-efficient signal processing using FPGAs. In: *Proceedings of the 2003 ACM/SIGDA Eleventh International Symposium on Field Programmable Gate Arrays*, Monterey, CA, 225–234. ACM.

21 Guo, Z., Najjar, W., Vahid, F., and Vissers, K. (2004). A quantitative analysis of the speedup factors of FPGAs over processors. In: *Proceedings of the 2004 ACM/SIGDA 12th international symposium on Field programmable gate arrays*, Monterey, 162–170. [s. n.], ACM.

22 Morales, J., Santiago, N., Leeser, M., and Fernandez, A. (2006). Hardware implementation of image space reconstruction algorithm using FPGAs. In: *49th IEEE International Midwest Symposium on Circuits and Systems*, 433–436. IEEE.

23 Tran, D., Chien, S., Sherwood, R. et al. (2004). DEMO: the autonomous sciencecraft experiment onboard the EO-1 spacecraft. In: *National Conference on Artificial Intelligence*, 1216–1217. AAAI Press.

24 Valencia, D., Plaza, A., Vega-Rodríguez, M.A., and Pérez, R.M. (2005). FPGA design and implementation of a fast pixel purity index algorithm for endmember extraction in hyperspectral imagery. In: *Chemical and Biological Standoff Detection III*, 599508-1–599508-10. SPIE.

25 Cai, T.T. and Silverman, B.W. (2001). Incorporating information on neighboring coefficients into wavelet estimation. *Sankhya: The Indian Journal of Statistic* 63: P127–P148.

26 Chen, G.Y., Bui, T.D., and Krzyak, A. (2005). Image de-noising using neighboring wavelet coefficients. *Integrated Computer-Aided Engineering* 12: P99–P107.

27 Kittisuwan, P. and Asdornwised, W. (2008). Wavelet-based image de-noising using neigh-shrink and bi-shrink threshold functions. In: *2008 5th International Conference on Electrical Engineering/Electronics, Computer, Telecommunications and Information Technology*, 497–500. IEEE.

28 Campenon, P. (2009). SPOT 6/7: continuity of SPOT 5 services in high resolution. In: *Beijing: the Sixth International Symposium on Digital Earth ISDE*.

29 Barnsley, M.J., Settle, J.J., Cutter, M.A. et al. (2004). The PROBA/CHRIS mission: a low-cost Smallsat for hyperspectral multiangle observations of the earth surface and atmosphere. *IEEE Transactions on Geoscience and Remote Sensing* 42 (7): 1512–1520.

30 Cutter, M.A. (2002). *CHRIS Data Format*, 1–31. London, UK: SIRA Electro-Optics.

31 Du, H. and Qi, H. (2004). An FPGA implementation of parallel ICA for dimensionality reduction in hyperspectral images. In: *IEEE International Geoscience and Remote Sensing Symposium*, Anchorage, 3257–3260. IEEE.

32 Miguel, A.C., Askew, A.R., Chang, A. et al. (2004). Reduced complexity wavelet-based predictive coding of hyperspectral images for FPGA implementation. In: *The Data Compression Conference (DCC' 04)*, 1–10. Snowbird, UT, USA: IEEE.

33 Vladimirova, T. (2004). *ChipSat–a System-on-a-Chip for Small Satellite Data Processing and Control Architectural Study and FPGA Implementation*. Surrey Satellite Technology Ltd.

7

Intelligent Communication Satellite System

7.1 Requirements for AI System Technology by Communication Satellite

Based on full investigation, in this section, we comprehensively consider the actual needs of models and future research directions and sort out the main requirements of current communication satellites for AI system technology. These requirements are not only the actual needs faced by communication satellites but also can be improved and gradually met through the development of AI systems and technologies. The research idea of this section is to sort out the requirements from three aspects: the intelligent autonomous management of satellite platforms, the intelligent distribution of loads, and the construction of low-orbit mobile Internet constellation, and form three main requirements: the platform's independent decision-making ability is limited, and the satellite security cannot be fully guaranteed. The load capacity distribution mode is solidified and affects the ease of use of the satellite. The need for autonomous operation of satellites in the construction of LEO mobile Internet constellation [1].

7.1.1 Satellite Security Requirements

At present, satellites, including communication satellites, generally have a limited level of intelligent decision-making, which leads to the most direct problem is that it is impossible to ensure the complete intelligent identification and recovery of faults, which will lead to missing the best time for fault handling when the satellite fails. However, the communication satellite system is becoming more and more complex, and the cost is also rising. Once the in-orbit fault is not handled in

Intelligent Satellite Design and Implementation, First Edition. Jianjun Zhang and Jing Li.
© 2024 The Institute of Electrical and Electronics Engineers, Inc. Published 2024 by John Wiley & Sons, Inc.

time, it will probably bring extremely serious economic losses and adverse social and even political impact.

Two typical satellite failure cases in orbit are introduced and analyzed below. These two cases are caused by the insufficient intelligence level of satellite systems and the lack of necessary independent decision-making and judgment, reflecting the problem that satellites cannot fully guarantee their own safety, which urgently needs to be improved by the research of artificial intelligence systems and technologies [2].

7.1.1.1 The Disintegration of Japan's Astro-H Satellite in Orbit

1) **Overview**: In 2016, the Japanese X-ray astronomical satellite was launched less than two months ago, and the reverse sign of the ground command caused high-speed rotation and disintegration, resulting in at least 1.8 billion yuan of economic losses.

2) **Process**: The X-ray astronomical satellite "Tong," jointly developed by JAXA and NASA, was launched on February 17, 2016, and the communication between the satellite and the ground was interrupted on March 26. Later, on March 27, the Space Joint Operations Command Center of the United States Aerospace Command released a message saying that at least five objects were found near the orbit of the "Astronomy-H" satellite, which was debris after the satellite was disintegrated.

3) **Influence**: The economic loss is at least 31 billion yen (about 1.888 billion yuan), which is "the tragedy of the scientific community."

4) **Cause**: "Tong" satellite appeared in abnormal conditions of the star sensor several weeks after launch. Whenever it ran over the abnormal area of the South Atlantic, the sensor would fail due to the influence of radiation anomaly. Due to the discovery of abnormal satellite attitude and orbit control, the ground injected the adjustment command in advance, but the positive and negative signs of the command were reversed. When the satellite executed the preset command, the thruster ignited in the wrong direction, resulting in a series of irreversible consequences for the attitude control system and the satellite, and the satellite rotated at high-speed and finally disintegrated.

5) **Reflect**
   - The system has weak decision-making ability and lacks basic error-handling mechanisms such as numerical detection.
   - The satellite was in an area not covered by the TT&C network at the time of the accident, and it was at midnight at the time of the accident, so the ground could not find and deal with it in time. This also shows that satellite safety cannot be fully guaranteed only by the whole-process monitoring on the ground [3].

7.1.1.2 Anik-F2 Satellite Communication Service Interruption

1) **Overview**: The communication service of the Anik-F2 satellite is interrupted due to software failure.

2) **Process**: Anik-F2 is a communication satellite developed by Boeing and operated by Telesat Company of Canada, which was launched into orbit on July 18, 2004. At 10:36 a.m. on October 6, 2011, the Anik-F2 satellite had technical problems and an abnormal load.

3) **Influence**: The application service was interrupted for several hours.

4) **Cause**: This anomaly is due to software design error and lack of sufficient judgment and decision. In a routine attitude maneuver, the satellite was mistakenly triggered to enter the safe mode, so the satellite turned off the load and controlled its orientation to the sun to ensure energy supply.

In conclusion, if the satellite has a certain level of intelligent decision-making, it can independently find and adjust the positive and negative signs of the command in a timely manner, or can independently avoid certain wrong logic, then these two accidents and the serious losses caused by them can be completely avoided. This also highlights the urgency of using the research of artificial intelligence systems and technologies to improve the intelligence level and independent decision-making ability of satellite platforms [4–6].

7.1.2 Load Usability Requirements

Now, communication satellites and their applications are influencing all walks of life with their extensive penetration and unparalleled progressiveness, and they are also an important means to solve globalization and connectivity and are receiving more and more attention from countries around the world. At the same time, the complexity of communication satellites is increasing. The focus of competition in the field of communication satellites has shifted from the success or failure of launch to the competition of performance indicators and continuous service capabilities. The requirements for the ease of use and usability of satellites are increasing. At present, the load of communication satellite cannot be flexibly adjusted. There is a problem that the load software and hardware are solidified, and the beam capacity distribution mode is solidified. This not only increases the cost of satellite operators, reduces the revenue of operators, reduces the user experience of satellite communication users, and affects the ease of use and usability of satellites [7–9].

On the one hand, the increase in the load capacity and the expansion of the functions of modern communication satellite platforms have put forward higher requirements for satellite payloads. The current communication satellite payloads generally face the problem that the hardware and software cannot be upgraded during the effective life of the satellite, which leads to the need to launch more advanced satellites to meet the continuous development of communication needs, greatly increasing costs.

On the other hand, the utilization rate of communication satellite capacity is closely related to the revenue of operators, which depends on whether the distribution mode of load capacity is reasonable. However, the satellite will conduct in-depth market research during its design to avoid insufficient and idle beam capacity, which will cause serious waste of onboard resources and affect user experience. However, from the current actual situation, the preliminary research cannot solve this problem well. The uneven distribution of load beam capacity is common in satellite projects of major operators. For example, the Ka-sat satellite in Europe and Viasat and Hughes in the United States have the problem of uneven beam heating and cooling [10–13].

In conclusion, the demand is clear and needs to be satisfied by the development of systems and technologies such as flexible payloads and intelligent load planning.

7.1.3 Requirements for Autonomous Operation of Satellites

For the information security facing global services, the LEO mobile constellation is the most important solution. It is the information network infrastructure to ensure "where the national interests extend, where the information network covers." It can achieve communication coverage in a large area and support the comprehensive application of space information in key areas such as national energy and resources development, environmental disaster reduction, infrastructure construction, emergency safety, marine transportation, fisheries, and other marine resources development, marine scientific research, and so on. It provides a safe and reliable information channel with the integration of space and earth, win-win cooperation for the interconnection of resource information, cultural communication information, business information, financial information, and public security information.

The construction of a low-orbit mobile Internet constellation can meet the overall development goal of both national strategic needs and commercial applications and can provide mobile communication, broadband communication, aviation navigation surveillance, and navigation enhancement services. The satellite system needs to have high-speed information processing capability, be able to provide real-time information network services for the platform and payload, and be able to realize data sharing. It is required to provide standard and flexible system services and adaptive communication protocols. It needs to have intelligent, independent management ability, with anti-interference, anti-destruction, independent management and independent survival ability, and can realize independent energy management, thermal control management, load management, network control management, fault diagnosis, and recovery under the condition of no intervention, so as to reduce the pressure of long-term management on the ground. These requirements cannot be fully met by existing technical means. It is urgent to use the research of artificial intelligence systems and technologies to meet the requirements of model tasks.

To sum up the above, in the field of communication satellites, the demand for AI system technology is mainly reflected in three major aspects: (i) Based on AI system technology improve the autonomous decision-making ability of satellite platforms to better ensure the safety of satellites. (ii) Based on artificial intelligence system technology, improve the optimization distribution and independent planning capacity of load capacity, improve the utilization rate of load capacity, and better meet the needs of users. (iii) Based on artificial intelligence system technology, improve the ability of independent operation and function management of single satellite, meet the requirements of autonomous multi-satellite networking and task coordination, and meet the requirements of low-orbit mobile Internet constellation construction.

7.2 Key Technologies of Communication Satellite Intelligent System

Intelligence is undoubtedly the future development direction of communication satellites. In the intelligent system of communication satellites, a series of key technologies are involved. This section will systematically elaborate on these key technologies.

7.2.1 Spectrum Sensing Technology in Satellite Communication

7.2.1.1 Introduction to Cognitive Radio

The concept of cognitive radio (CR) was clearly put forward by the consultant of MITRE, Dr. Joseph Mitola of the Royal Swedish Institute of Technology, and Professor GERALD Q MAGUIRE, JR in the IEEE Personal Communication magazine in August 1999. It believes that CR is an intelligent wireless communication system, which can sense the surrounding communication environment, adaptively adjust the internal communication mechanism through learning the changes in the surrounding environment, to adapt to the changes in the external environment, improve the stability of communication, and improve spectrum utilization. CR improves the flexibility of personal wireless services through "Radio Knowledge Representation Language (RKRI)," which was then discussed in detail in the doctoral thesis defense held by the Royal Swedish Academy of Sciences in 2000.

Since Dr. Mitola first proposed the concept of CR and systematically expounded the principle of CR, different institutions and scholars have given the definition of CR from different perspectives, among which the more representative ones are the definition of FCC (Federal Communication Commission) and Professor Simon Haykin. FCC believes that "CR is a radio that can change transmitter parameters based on the interaction of its working environment." Professor Simon Haykin believes that CR is an intelligent wireless communication system. It can sense the external environment and use artificial intelligence technology to learn from the environment. By changing some operating parameters (such as transmission power, carrier frequency, and modulation technology) in real time, it can adapt its internal state to the statistical changes of the received wireless signals, so as to achieve the following goals: highly reliable communication at anytime and anywhere: effective use of spectrum resources [14].

Summing up the above definitions, we can find that CR should first have cognitive ability, which enables CR to capture or perceive information from the wireless environment in which it works, so as to identify unused spectrum resources (i.e. spectrum holes) in specific time and space, and select the most appropriate spectrum and operating parameters. This task includes three main steps: spectrum sensing, spectrum analysis, and spectrum decision. The main function of spectrum sensing is to monitor available frequency bands and detect spectrum holes. Spectrum analysis estimates the characteristics of spectrum holes obtained by spectrum sensing. Spectrum decision selects the appropriate frequency band to transmit data according to the characteristics of spectrum hole and user needs. Secondly, CR should have adaptive ability, which enables CR equipment to dynamically program according to the wireless environment, thus allowing CR equipment to use different wireless transmission technologies to transmit and receive data. The parameters that can be reconstructed include: operating frequency, modulation mode, transmission power, and communication protocol. The core idea of adaptation is to provide reliable communication services using the free spectrum of the authorized system without harmful interference to the spectrum authorized user (LU). Once the frequency band is used by LU, CR has two ways to deal with it: one is to switch to other idle frequency bands for communication. Second, continue to use this frequency band, but change the transmission statistics or modulation scheme to avoid harmful interference to LU.

CR is essentially an extension of software radio. It uses the method based on the radio domain model to process the rules that control the use of radio spectrum, so as to describe radio rules, equipment, software modules, radio wave transmission characteristics, user needs, networks, etc. To increase the flexibility of radio spectrum resource utilization, so that traditional software radio can better serve users. The system model is described in Figure 7.1.

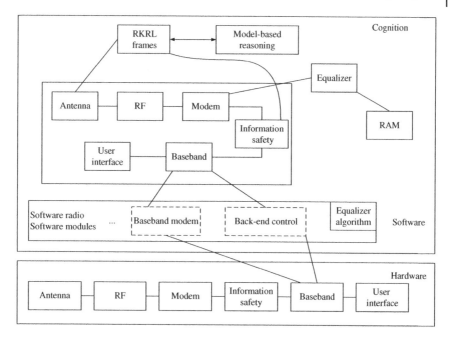

Figure 7.1 Conceptual model of Mitola.

The practical significance of CR is that it can improve user capacity and spectrum efficiency and solve the problem of spectrum resource waste. CR solves the contradiction between the demand of wireless users for spectrum resources and the insufficient utilization of classified spectrum resources. In addition, there are intelligent features in the advanced business layer of the communication system, such as the use of optimization methods to achieve the mutual transmission of information in the limited signal space, thus opening a new space for efficiency, coexistence, compatibility, and interoperability of radio communication users in the increasingly crowded wireless spectrum resources. On December 30, 2003, the FCC of the United States issued a report on soliciting telecommunication policy reform plans (NPRM), including CR technology and related applications. According to the announcement, "wireless terminals with CR function can use a certain frequency band according to their own needs as long as they find that a certain frequency band is not used in the frequency domain or airspace, even if they have not obtained the spectrum use license," which will be one of the application prospects of CR. For this reason, FCC has opened the 5 GHz U-NII bandwidth spectrum and 7.5 GHz bandwidth for ultra-wideband signals in the range of 3.1–10.6 GHz. At present, CR has attracted more and more scholars' attention [15].

Because the spectrum access priority of cognitive users is lower than that of primary users, cognitive users must independently detect whether the spectrum is idle and whether the primary user appears, and can also ensure the efficiency of spectrum use without interference to the primary user. The operation behavior of cognitive users is related to their interaction with the surrounding environment. A CR user can sense whether a frequency band is being used. If it is not used, it can occupy this frequency band without causing harmful interference to the relevant main user. If the primary user of this frequency band restarts the transmission later, the CR user can jump to another frequency band or avoid interference with the primary user by changing its transmission power level and or modulation mechanism, so as to continue to use this frequency band. It can be seen from this that the primary task of CR is spectrum sensing. Therefore, in the research, a key problem to be solved is how to make CR systems have sensitive spectrum sensing ability to capture and handle the power and performance problems involved in the signal of a specific frequency band in different wireless communication environments. It is required to adopt advanced signal processing technology and propose spectrum sensing algorithm with strong adaptability, good real-time performance, and high accuracy [16].

7.2.1.2 Spectrum Parameters

The effective use of frequency band by sensing technology requires the system designer and specification maker to define the parameters of the optional transmission mode of the system. Too few parameter choices will limit the ability of perception and adaptation. Too many parameter choices will increase the complexity of the system. A new concept of spectrum parameters was proposed in the report of the Spectrum Regulatory Committee: "The spectrum regulatory committee has analyzed the benefits of using frequency, power, space and time to represent spectrum resources. In the past, the committee has realized that the first three variables of spectrum resources are represented, and only the first three variables are considered when spectrum is authorized for use. With the emergence and development of new technologies, the committee has considered adding time variables as the variables of spectrum resources, which is more conducive to dynamic spectrum allocation and assignment of special spectrum usage rights."

The new spectrum resource parameters are defined on the basis of the original frequency, space, and energy by adding one dimension of time, so that the spectrum usage rights can be specified and allocated more dynamically. The existing intelligent sensing system is developed based on this definition.

The definition of spectrum resource parameters is to show that spectrum sensing technology can adjust all parameters in spectrum resources to make the use of frequency bands more flexible. The latest digital signal processing technology and antenna technology can realize spectrum sensing in the entire spectrum space [17].

For satellite communication, the power parameter of the spectrum can be used as the control variable of the transmitter's transmission power, the frequency parameter is used as the modulation frequency of the transmission antenna, the time parameter can provide real-time reference for the second user's spectrum conversion, and the spatial parameter of the spectrum determines which satellite the second user needs to aim at for communication. To sum up, these spectrum parameters must be calculated by the sensing station in the satellite cognitive network, and stored as various elements of a spectrum vector in the alternative spectrum resource table.

7.2.1.3 Spectrum Sensing Concept

The spectrum sensing technology has the following two aspects:

1) Through automatic perception of the changing radio electromagnetic environment, signals are transmitted on the frequency band where no other system works. First, the idle frequency band that is not occupied by other systems is sensed and used to transmit information. This intelligent sensing system can be used in cooperation with other systems to improve the efficiency of spectrum utilization. It can also share the information of the spectrum environment with other similar devices to ensure that there is no interactive interference. However, the system is realized on the premise that the frequency in the spectrum space is not occupied for a long time. With the constant change of time and frequency, the available spectrum resources of the sensing system are also changing [18].

2) Efficient use of spectrum through sharing information and operation. For example, between different users and subsystems, the same channel is shared by TDMA or CDMA coding. This method can achieve adaptive and dynamic operation through network cooperation. By sharing various spectrum parameters, it can greatly improve the utilization of spectrum.

At present, the main technical problems of spectrum sensing lie in the following two aspects:

1) Cognitive technology of electromagnetic environment and adaptive operation under this cognitive condition.

2) The influence of the combination of various spectrum parameters on adaptive technology.

The main advantages of using spectrum sensing technology in satellite communication networks are as follows:

1) Improve spectrum utilization.

2) Maintain service quality under complex and changeable external conditions.

3) Adjust the transmission to reduce interference to other systems.

To sum up, the spectrum sensing technology can lead to the following two problems:

1) Perceived idle frequency band.
2) Optimize the use of spectrum resources according to the sensing results.

The idle frequency band refers to the spectrum resources that are not used by the main user. The designation of idle frequency band includes spectrum sensing and prediction of its usage trend. Therefore, understanding the rules used by the main user in this frequency band can effectively improve the accuracy of this assignment. The usage rules of the frequency band can be stored in the cognitive system in advance by downloading or writing to the memory [19].

7.2.1.4 Spectrum Sensing Algorithm

According to the elaboration of the basic concept of CR, it needs to have the ability to sense the surrounding environment in a wide range of frequencies, so as to provide users with the ability to meet the communication needs to the maximum extent. This requires CR equipment to accurately sense whether there is an idle frequency band at a certain time and place for the second user to use. At the same time, it is also necessary to monitor whether there are new legal users who need to access the frequency band at any time so that the second user can withdraw from using the spectrum resources in time and avoid interference to the primary user. The spectrum sensing technology to detect whether a certain frequency band is occupied by PU (primary user) plays an important role in the realization of CR. With the continuous development of CR, spectrum sensing technology will have more and more extensive application prospects.

Therefore, when analyzing the spectrum sensing technology of CR, we should first consider how to determine whether a certain frequency band is idle in the most reliable way, and ensure that the detection time is short enough to enable the second user to access and exit in real time.

At present, there are many methods in the field of CR to detect whether there are signals in a certain frequency band. According to the detection type, it can be divided into two types: signal presence detection and signal coverage detection. According to the number of detection nodes, it can be divided into single-node detection and multi-node joint detection. According to the detection methods, it is mainly divided into three categories: matched filtering, energy detection, and periodic characteristic detection.

7.2.2 Intelligent Information Distribution and Push Technology

At present, the distribution mode of satellite communication system is basically unidirectional push, and there are few "personalized" designs of resources based on the differences of terminals. Therefore, the use of publish/subscribe mode makes the resources and end users loosely coupled, and separate the two, so that different resources can be conveniently distributed to different terminals. Therefore, resource multicast and resource subscription push functions for specific terminal groups can be realized. Due to the great difference between satellite communication and ground network, it is necessary to study the publish and subscribe working mode in satellite systems and design it according to the characteristics of satellite communication [20].

Due to the extremely limited storage space and computing capacity on the satellite, it is impossible to copy the publish/subscribe paradigm on the ground to provide resource services on the satellite. It is necessary to carry out targeted protocol design according to the characteristics of satellite communication. A typical publish/subscribe system includes topology, event model, subscription model, matching algorithm, routing algorithm, and quality of service assurance. The event model defines the data structure of the event, the subscription model defines the subscription conditions that the system can support, and indicates how subscribers express their interest in the event subset. The event model and subscription model jointly determine the expression ability of the system. The matching algorithm is generally optimized by combining the corresponding event and subscription models. The topology determines the scalability of the system. The routing algorithm is generally optimized according to the topology of the corresponding event agent network. Considering the characteristics of the communication satellite system, the topology is a satellite-centered star, and the route is a single-hop connection, so the key technologies that need to be designed include: (i) the event and subscription model, including how to express the event, the producer publishes the resource, the consumer subscribes to the data, and the notification service notifies the consumer of the generation of new subscription resources. (ii) Data flow design, including consumer resource request, notification service forwarding request to producer, producer uploading resource data, notification service forwarding resource data to consumer, subscriber canceling subscription, etc. (iii) The resource matching algorithm, how to match user subscriptions with resources, is the most critical technology in the publish/subscribe system, which determines the working efficiency of the system [21].

7.2.3 Satellite Digital Channelization Technology

7.2.3.1 Basic Structure of Onboard Flexible Transponder

The flexible repeater is a nonregenerative transparent repeater satellite that is suitable for broadband nonuniform bandwidth signal transmission and can be used for onboard digital signal processing, based on the concept of the United States Broadband Global Satellite System (WGS) and digital channelization

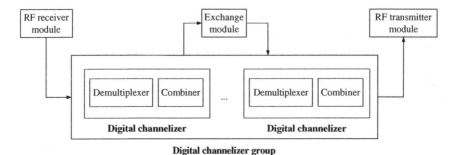

Figure 7.2 Structure diagram of digital channelizer.

technology, applying broadband filter bank theory, multi-sampling rate digital signal processing theory, polyphase decomposition method of filter bank, software radio implementation method, etc. It is one of the payloads of the future broadband satellite system. This transponder and its related technologies can be applied to broadband mobile satellite communication system, mobile communication system, and other fields. Structure diagram of digital channelizer is shown in Figure 7.2.

In Figure 7.2, RF receiver module includes down-conversion, low-noise amplifier, and other modules. RF transmission module includes up-conversion, power amplification and other modules. The structure in the solid line frame is a digital channelizer group, including a plurality of digital channelizers. The structure in the dashed frame is a digital channelizer, including a splitter and a combiner. The switching module is a circuit switching matrix, which supports real-time exchange between multiple beam coverage areas and multiple user sample data beams in each coverage area.

The digital channelizer is one of the basic components of the flexible transponder, which realizes the function of demultiplexing the received signals in the specific beam coverage area and combining the exchanged signals. It realizes the exchange of all user signals in each beam coverage area by combining with the exchange module.

The WGS system of the US military uses digital channelizer technology. The bandwidth of each broadband uplink and downlink channel is 125 MHz. Each broadband uplink and downlink channel is divided into 48 basic subchannels. Any service subsignal can occupy one or several adjacent basic subchannels. There is a protection band between different service subsignals. All service subsignals have switching/routing functions as shown in the Figure 7.3.

It can be seen from Figure 7.3 that the implementation steps of the digital channelizer include three steps: (i) Separate and extract each service subsignal in the broadband uplink channel; (ii) exchange the separated service subsignals

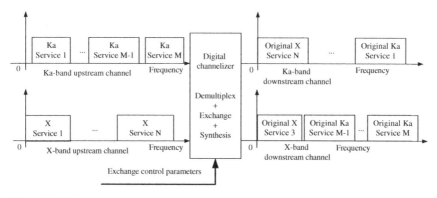

Figure 7.3 Schematic diagram of WGS digital channelizer.

according to the exchange control parameters; and (iii) after switching, each service subsignal is integrated to form a broadband downlink channel signal.

7.2.3.2 Digital Channeler Signal Separation and Signal Synthesis Algorithm

At present, the implementation methods of step (i) and step (iii) include digital up-down-conversion method, analytical signal method, polyphase discrete Fourier transform method and multi-level method. These methods almost assume that the broadband uplink channel is uniformly divided into several traffic subchannels with the same bandwidth. Therefore, none of the above methods can meet the requirements of nonuniform bandwidth traffic subchannels division.

It is speculated that the WGS implementation method using digital channelizer should be suitable for the application scenario where the traffic subchannels are divided evenly or unevenly, and each broadband uplink, and downlink channel contains hundreds of traffic subsignals.

Time Domain Discrete Filter Bank To realize the separation, exchange, and postexchange synthesis of nonuniform bandwidth traffic subsignals, the traditional method uses the discrete filter bank (DFB) method in the time domain. In this channelization method, separate filter banks and integrated filter banks need to be designed for specific traffic subchannel division scheme to complete the separation/exchange of various traffic subsignals and the postexchange integration of various traffic subsignals.

The signal separation unit determines each digital local oscillator frequency according to the center frequency of each service subsignal in a broadband uplink channel complex baseband signal, determines each equiripple low-pass filter coefficient and each extraction factor according to the bandwidth of each service subsignal, and separates the baseband signal of each service subsignal. Then, according to the exchange control parameters, the baseband signals of all services are exchanged. According to the baseband signal bandwidth of each service in a certain downlink channel after exchange and the spectrum position occupied by the downlink channel, interpolation filtering, spectrum shift and addition are performed, and finally the broadband downlink channel signals are formed.

This method is simple, economical, and reliable when the number of traffic subsignals is small. However, when a channel contains dozens or even hundreds of traffic subsignals with nonuniform bandwidth, this method has little practical value in terms of flexibility and computational complexity of storage filter bank coefficients.

Frequency Domain Filtering Method Frequency domain filtering method is a filtering calculation method based on fast Fourier transform and inverse fast Fourier transform. The basic implementation method is shown in Figure 7.4.

The digital channelizer based on the frequency domain filtering method is suitable for both uniform bandwidth traffic subsignal applications and nonuniform bandwidth traffic subsignal applications. The digital channelizer based on the frequency domain filtering method has three advantages over the digital channelizer based on the time domain method:

1) **High flexibility**: Because the number of service subsignals contained in the second broadband uplink channel is flexible and the center frequency and

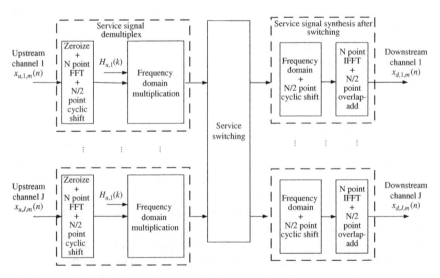

Figure 7.4 Implementation block diagram of digital channelizer based on frequency domain filtering method.

bandwidth of each service subsignal are variable, if the time domain method is used, a certain uplink channel separation unit can only configure the spectrum shift, low-pass filtering, and extraction processing branches in parallel according to the maximum possible number of service subsignals, which is difficult to change flexibly according to the number of service subsignals, resulting in resource waste. The broadband downlink channel synthesis unit also faces the same problem. Using the frequency domain filtering method, it is only necessary to configure the values of the elements in the corresponding frequency domain filter parameter array online according to the number of service subsignals contained in an uplink channel and the center frequency and bandwidth of each service subsignal. Other processing remains unchanged. This can save processing resources and have high flexibility.

2) **Low reserve**: The channel separation unit needs much more space to store the filter coefficients of the time-domain method than the filter parameters of the frequency-domain method, which can reduce the consumption of hardware resources.

3) **Low computational complexity**: Each broadband uplink channel separation unit based on the frequency domain filtering method needs only one N-point FFT operation and N-point complex multiplication operation to complete the separation function. Compared with the time domain DFB method, the computational complexity is greatly reduced.

7.2.4 Spaceborne Intelligent Antenna Beamforming Technology

7.2.4.1 Classification of Intelligent Antenna Digital Beamforming Technology

At present, the digital beamforming technology of intelligent antenna can be divided into three categories: beamforming method based on reference signal, beamforming method based on direction of arrival (DOA), and blind beamforming algorithm based on signal structure characteristics, and the most widely used are beamforming method based on direction of arrival (DOA) and beamforming method based on reference signal. The digital beamforming method of satellite intelligent antenna, based on ensuring the performance of the algorithm, puts forward higher requirements for the anti-interference ability and computational complexity of the algorithm, and the algorithm should be suitable for hardware implementation. Least mean square (LMS) is one of the most widely used beamforming methods for intelligent antennas. On the one hand, LMS algorithm has less computation and is easy to implement by hardware, On the other hand, LMS algorithm can adaptively lock and track the required signal and, at the same time, align the zero-response point of the antenna pattern with the jamming signal, so it has a good effect in anti-interference signal.

7.2.4.2 Broadband Digital Beamforming Technology

With the increasing demand for communication services and the real-time requirements of the battlefield, the original single telephone and narrowband data are gradually developing toward multimedia services such as voice, image, text, video, and data, which requires satellite communications to achieve broadband data transmission to meet the needs of new services. Broadband data service will be one of the important development directions of satellite communication. At present, the available spectrum resources of satellite communications are very limited, and broadband data transmission is bound to develop toward higher frequencies. Because Ka-band has the characteristics of wide available bandwidth, less interference, and small equipment size, Ka-band has been gradually used to realize broadband data transmission in recent years. Then, how to deal with broadband signals and how the broadband signals will affect the smart antenna will be another key problem to be solved.

For broadband signals, if the conventional weighted summation beamforming method is used, the beam patterns at each frequency are different, which will lead to distortion of the signal waveform incident from the direction beyond the maximum response axis of the beam and directly affect the system's waveform estimation and target characteristics identification and a series of functions. Therefore, the exploration of broadband signal processing technology will be of great significance. The basic method to solve this problem is to design a constant beamwidth beamformer so that the array system has the same beam pattern corresponding to the input signals of different frequencies to effectively protect the waveform of the input signals and provide high-quality distortion-free data for subsequent target recognition and other functions. At present, there are many methods of constant beamwidth beamforming, such as the Fourier transform method, Chebyshev weighting method, Bessel function method, least squares estimation method, etc. These methods can generally achieve the effect of constant beamwidth and can be realized by numerical calculation.

7.3 Typical Application Cases

7.3.1 Software Definition Satellite

In the traditional mode, the communication satellite is faced with the difficulties that the designed frequency and coverage scheme is difficult to adjust in orbit, and the size and transmission rate of the ground antenna need to be improved with high-cost adaptability. With the development of satellite communication application demand, the demand for satellite communication is becoming more and more diversified. For example, the demand for security protection capability, communication rate, and so on presents dynamic changes in time and space. To improve the flexibility, security, and reliability of satellite communication applications, it is urgent to carry out the research on software-defined satellites.

Adopt artificial intelligence technology, master the available satellites, available service providers, available modems, etc., of the current terminal, and realize the cooperative operation between different satellite communication technologies, so as to achieve efficient and optimized configuration. Software is used to realize reconfiguration, so that the load coverage, bandwidth, function-frequency, etc. can be upgraded with the change of demand.

The load will have intelligent forwarding capabilities such as allocating resources on demand, adjusting the system, and changing the switching route mode. Through satellite-ground joint learning, the payload has the ability of task reconfiguration for intelligent applications, including dynamically changing the functional configuration of antennas, transponders, and other payloads. It can intelligently adjust parameters such as service beam, working frequency band, bandwidth, power, and communication system according to user environment and requirements. The satellite can flexibly and dynamically meet the changes of user needs under different business backgrounds, saving the cost of the improvement.

7.3.2 Autonomous Orbit Change/Orbit Maneuver of Satellite

The traditional satellite relies on the ground command system to launch and change orbit, which has high launch cost and poor damage resistance. In a special period or war, once the ground system is destroyed, the satellite does not have the ability to launch, and cannot adapt to the mission characteristics of strong real-time and strong confrontation in future space operations. Therefore, the satellite platform has an urgent need for autonomous orbit change [22].

After the separation of the satellite and the rocket, the satellite will automatically conduct orbit change control. The satellite realizes its own orbit perception by fusing various sensor data, calculates and generates orbit change strategy independently according to the specified orbit position, and realizes real-time closed-loop control and correction during the satellite ignition, and finally reaches the specified position. During orbit change, the satellite realizes independent health management, and quickly identifies and processes abnormal faults. The intelligent technologies involved include (i) State self-awareness technology in complex environment, using multi-level and multi-source information fusion technology based on inertial measurement and dynamic model, astronomical navigation and multimode GNSS measurement and navigation to achieve fully autonomous navigation and autonomous orbit maintenance. (ii) The generation technology of satellite orbit change strategy is based on feedback control guidance and other methods to independently complete the transfer orbit design, optimize and calculate the orbit control strategy according to the dynamic model (or neural network type, etc.), objective function, and satellite resources, and complete the generation of orbit change strategy based on autonomous mission planning technology. During the orbit change, the satellite detects the engine thrust, specific impulse, tank pressure, temperature, ignition direction, and other telemetry parameters in real time, corrects and optimizes its orbit change strategy, and applies the optimized results to the actuator in real time to form closed-loop control. (iii) Satellite fault diagnosis technology based on deep learning. In order to realize the autonomous and healthy operation of satellite after orbit change and fixed point, the onboard application software is based on the neural network model after ground training to monitor the satellite status in real time, diagnose, and predict its working status, performance and index trend, and locate the fault in time when the fault occurs, and determine the components that are not working normally or whose performance is degraded [22].

7.3.3 Satellite Intelligent Spectrum Sensing and Anti-interference

At present, the spectrum sensing problem in satellite communication can be solved by existing signal processing technology. The cyclostationary detection is a hot topic in current research. Compared with energy detection, the disadvantage of this detection method is the high computational complexity. Therefore, it is necessary to use artificial intelligence to reduce the complexity of the algorithm and improve the spectrum sensing and anti-interference ability. At the same time, facing the future space networking satellites, users' demand for information has changed greatly. The traditional communication mechanism based on request/response mode is difficult to adapt to the needs of large-scale, asynchronous, and multi-point communication due to its tight coupling characteristics. The service application should change from a relatively closed, familiar user-oriented and relatively static tight coupling form to an open, publicly accessible and dynamic collaboration intelligent loose coupling mode [23].

It is an inevitable trend in the field of communication satellites in the future to develop the intelligent system of communication satellites and other related technologies to improve the processing capacity and intelligent level of space systems. The artificial intelligence technology is used to intelligently adjust frequency and power to meet the explosive frequency demand in the future. At the same time, the jamming signal spectrum database is established through intelligent spectrum sensing and interference detection, and the results of spectrum monitoring are used for dynamic channel management and interference avoidance, so as to realize the intelligent anti-jamming function of the satellite: (i) the spectrum sensing technology based on artificial intelligence is adopted, in satellite communication, it can sense the working status of the current target transponder, intelligently analyze the use of frequency and power, and adjust the parameters in the spectrum pool in real-time through the understanding and learning of the second user, to meet the flexible configuration requirements of spectrum resources for future space networking. (ii) The intelligent antenna technology based on digital beam forming is used to dynamically form the required beams in the digital domain and establish a real-time jamming signal spectrum database. When the antenna is in the receiving state, it can ensure that the gain in the desired signal direction is not affected, and at the same time, it can adaptively align the zero point of the pattern with the direction of the interference signal to play the role of suppressing interference. When it is in the transmitting state, the main lobe gain of the generated beam is higher and the side lobe is lower, which reduces the probability of the communication signal being intercepted by the enemy [24–26].

Through real-time monitoring and analysis, we can accurately know the dynamic changes of the interference situation of nonpartners, achieve reliable data transmission, and improve user experience.

References

1 Shishko, R., Aster, R., Chamberlain, R.G. et al. (1995). NASA systems engineering handbook. NASA Sti/recon Technical Report N, 6105.

2 Martin, L. (2016). *ORBCOMM to Explore Better Connected Machines*. ORBCOMM Inc.

3 Osorio, R.V. and Lemos, J.P. (2006). Scos-2000 Release 4.0: multi-mission multi-domain capabilities in ESA SCOS-2000 MCS kernel. In: *2006 IEEE Aerospace Conference*, 1–17. IEEE.

4 Chamoun, J.P., Risner, S., Beech, T. et al. (2006). Bridging ESA and NASA Worlds: lessons learned from the integration of hifly®/SCOS-2000 in NASA's GMSEC. In: *2006 IEEE Aerospace Conference*, 1–8. IEEE.

5 Gudmundsson, V., Schulze, C., Ganesan, D. et al. (2015). Model-based testing of NASA's GMSEC, a reusable framework for ground system software. *Innovations in Systems and Software Engineering* 11 (3): 217–232.

6 Hrobak, M. (2015). Synthetic instruments. In: *Critical mm-Wave Components for Synthetic Automatic Test Systems* (ed. M. Hrobak), 1–10. Wiesbaden: Springer Vieweg.

7 Niezette, M. and Lucia, D. (2018). An EGS-CC-based core control segment. In: *15th International Conference on Space Operations*, 2684. SpaceOps.

8 Widegård, K. and Pecchioli, M. (2018). Harmonisation of products for ground segment operations at ESA. In: *2018 SpaceOps Conference*, 2626. SpaceOps.

9 TeKolste, R.D. and Liu, V.K. (2018). Outcoupling grating for augmented reality system. US Patent Application 10/073,267. 11 September 2018.

10 Chakravarthula, P., Dunn, D., Akşit, K. et al. (2018). FocusAR: auto-focus augmented reality eyeglasses for both real and virtual. *IEEE Transactions on Visualization and Computer Graphics* 24 (11): 2906–2916.

11 CECIMO (2017). *European Additive Manufacturing Strategy*. Filip Geerts Editor Vincenzo Renda.

12 Wohlers Associates (2017). Wohlers 2017 Report on 3D printing industry points to softened growth. Wohlers Report 2017.

13 CECIMO (2017). Economic and statistical toolbox. European Association of the Machine Tool Industries and related Manufacturing Technologies. January 2017.

14 Zhu, G.X., Li, D.C., and Zhang, A.F. (2012). The influence of laser and powder defocusing characteristics on the surface quality in laser direct metal deposition. *Optics and Laser Technology* 44 (2): 349–356.

15 Sarah Loff (2016). Researching 3D Printing Technology on the Space Station. https://www.nasa.gov/image-feature/researching-3d-printing-technology-on-the-space-station.

16 Michael Peck (2017). Air Force satellites will use 3-D printed parts. https://www.c4isrnet.com/c2-comms/satellites/2017/04/13/air-force-satellites-will-use-3-d-printed-parts/

17 The European Space Agency (2017). Metal 3D-printed waveguides proven for telecom satellites. https://www.esa.int/Enabling_Support/Space_Engineering_Technology/Metal_3D-printed_waveguides_proven_for_telecom_satellites.

18 International Federation of Robotics (IFR) (2016). European Union occupies top position in the global automation race. World Robotics Report 2016. 29 September 2016.

19 International Federation of Robotics (IFR) (2017). Deployment of robots soars 70 percent in Asia. FRANKFURT, Germany. 01 Feburary 2017.

20 International Telecommunication Union (2005). *Internet Reports 2005: The Internet of Things*. Geneva: ITU.

21 NYIT (2017). X-wave Innovations Team Up to Develop RFID Technology for NASA. http://www.rfidjournal.com/articles/view16426/1 (accessed September 2017).

22 Digi International takes IoT into space with latest NASA program http://www.mwee.com/news/digi-international-takes-iot-space, 05 May 2017.

23 DCU Alpha (2017). ESA look to the stars for IoT projects – TechCentral. http://www.techcentral.ie/dcu-alpha-esa-look-stars-iot-projects/.

24 Sowmya, R. and Suneetha, K.R. (2017). Data mining with big data. In: *International Conference on Intelligent Systems and Control*, 246–250. IEEE.

25 Qian, D. (2016). High performance computing: a brief review and prospects. *National Science Review* 3 (1): 16.

26 Maseleno, A., Huda, M., Teh, K.S.M. et al. (2018). Understanding Modern Learning Environment (MLE) in big data era. *International Journal of Emerging Technologies in Learning (iJET)* 13 (05): 71–85.

8

Intelligent Navigation Satellite System

At present, the global satellite navigation systems that have been put into use or are under construction include GPS in the United States, GLONASS in Russia, GALILEO in Europe and Beidou in China, which can provide users with positioning, speed measurement, and time service. With the coming of a new wave of artificial intelligence, the artificial intelligence system technology will make valuable contributions to the improvement of the service performance of navigation satellites, the evolution of satellite functions, the expansion and improvement of constellation business, and finally, realize the intelligent, autonomous, flexible, and stable satellite system. At present, the development of mainstream navigation satellites has gradually moved toward intelligence. Taking GPS-IIIF satellite as an example, its distinctive feature is the full digital payload. Compared with the 70% digital degree of the GPS-IIIA satellite launched in 2018, the GPS-IIIF satellite has achieved a full digital payload, which lays a solid foundation for smart and intelligent navigation services in the future.

Based on the composition analysis of navigation satellite artificial intelligence system technology, relevant technologies can be divided into constellation network intelligent management, navigation task intelligent management, satellite autonomous health management, and satellite intelligent on-orbit maintenance according to the characteristics of navigation business and navigation satellite, as shown in Figure 8.1. The first two technologies focus on the business level of satellite system, while the latter two technologies focus on the basic support level of satellite [1].

Among them, constellation network intelligent management aims at the link and network management, system topology and routing, network resource application configuration, constellation-level autonomous navigation, etc., of navigation satellite constellation network. The intelligent management of navigation tasks aims at satellite load service, positioning, and navigation timing service function and performance. Satellite autonomous health management aims at the

Intelligent Satellite Design and Implementation, First Edition. Jianjun Zhang and Jing Li.
© 2024 The Institute of Electrical and Electronics Engineers, Inc. Published 2024 by John Wiley & Sons, Inc.

Figure 8.1 Technical composition analysis framework.

health status assessment and fault management required for the continuous, stable, and reliable on-orbit operation of navigation satellite systems. Satellite intelligent on-orbit maintenance aims at the on-orbit autonomous evolution, autonomous repair and environmental adaptation of the software and hardware of the on-board electronic information system [2].

8.1 Intelligent Management of Constellation Network

8.1.1 Satellite Network Intelligent Management Technology

The intelligent space-based network with multiple functions, complementary orbits, high intelligence, autonomous operation, and easy expansion has become a new development direction. In addition, with the rapid development of terrestrial mobile communication networks, how to connect space-based and terrestrial networks and lay the foundation for seamless communication and integration of space-based networks has also become an important research direction. With the development of space technology, integrated information network of space and earth should provide conditions for different space network nodes to share resources and provide conditions for various satellites to access the space information network [3].

Satellite network intelligent management mainly involves key technologies such as network architecture, inter-satellite link communication, network topology, network routing, etc. The key of intelligent management technology of navigation constellation is intelligent networking and integrated communication access technology with ground or communication satellite network. Through the links between satellites, satellite-ground, starry sky and ground stations, users, aircraft, and various communication platforms in the ground, sea, air, and deep air are intensively combined, and intelligent high-speed processing, switching, and routing technologies are adopted to carry out accurate information acquisition, rapid processing, and efficient transmission of integrated information networks, namely, integrated space-based, air-based, and land-based networks.

8.1.2 Constellation-level Intelligent Autonomous Navigation Technology

Satellite autonomous navigation relies on the software and hardware equipment of the satellite itself to independently determine its position, speed, and attitude in space and provide satellite orbit and attitude control information when flying in space. It is a prerequisite for the satellite to realize intelligent autonomous control and high-performance autonomous flight, and it is also the trend in the development of satellite control technology today. It has great significance in reducing the burden of ground measurement and control, reducing the cost of satellite operation, improving the viability of satellites, and expanding the application potential of satellites. The autonomous operation requirements of satellite networking also put forward higher requirements for satellite autonomous navigation technology [4].

In addition, the magnitude and frequency of orbit control after the navigation satellite is put into orbit directly affect the navigation service performance. High-precision orbit autonomous maintenance technology strives to obtain the orbit position of the local satellite in real time through high-precision accelerometer and autonomous navigation technology and, at the same time, maintain the orbit in real time through micro-new small thruster, which can greatly reduce the orbit control frequency and extend the satellite navigation service time. Intelligent autonomous navigation technology includes the following means:

1) For earth-orbiting satellites, "infrared earth sensor + star sensor" is a relatively easy astronomical navigation method. Based on the measurement of these two sensors, the included angle between the star direction and the ground direction and the geocentric distance can be calculated, and the satellite orbit can be determined by further using modern estimation theory.

2) The geomagnetic sensor is used for geomagnetic navigation to measure the geomagnetic field intensity of the satellite's location and then match it with the International Geomagnetic Field Model (IGRF). The orbit information of the satellite is obtained by designing the corresponding filtering solution method. Using geomagnetic information for autonomous navigation, the required hardware structure is simple and easy to implement. However, the accuracy of geomagnetic navigation still cannot meet the requirements. The key of this technology is to estimate and correct the error of geomagnetic model by combining different measurement information, so as to continuously improve the accuracy of geomagnetic navigation autonomous orbit determination. Geomagnetic navigation uses the geomagnetic information inherent in the earth itself, with low cost and power consumption. It is mainly applied to ground vehicles (cruise missiles, aircraft, etc.) and can also have high accuracy for low-orbit satellites, which mainly depends on the accuracy of magnetometers. Considering the magnetic interference and radiation in space, the application of geomagnetic navigation is obviously limited for autonomous navigation of high-orbit satellites [5].

3) Direct sensitive horizon and indirect sensitive horizon using starlight refraction are used for autonomous celestial navigation based on star sensors. Star sensors are mainly sensitive to star information. Because the star's opening angle is very small, their position in the inertial coordinate system can be accurately known after a long time of astronomical observation and recording. Therefore, taking stars as the measurement object for navigation can have high accuracy, even reaching the angular second measurement accuracy. This is one order of magnitude higher than the sun sensor and 2 orders of magnitude higher than the general earth sensor. It has strong autonomy and is not affected by orbit. It is the attitude sensor with the highest accuracy among all satellite-borne sensors. Autonomous navigation technology based on star sensors has a wide range of applications. The method of directly sensing the horizon only needs to use star sensors and infrared horizon sensors. It has the advantages of easy access to observed values, mature technology, good reliability, and a simple algorithm. However, the infrared band characteristics of the earth radiation are not stable

enough, which leads to the low accuracy of the infrared horizon meter in measuring the horizon, and thus reduces the navigation accuracy of the satellite. The autonomous navigation method with indirect sensitive horizon can obtain high navigation accuracy by measuring the refraction of starlight, but this method is limited by the influence of the earth's atmospheric model. The earth's atmospheric model is also affected by seasons, latitudes, and solar activities, so there will be uncertain system errors in the observation model, which will lead to large navigation errors. The current research found that this method is difficult to further improve navigation accuracy [6].

4) Autonomous navigation based on ultraviolet sensors can observe multiple celestial object information in multiple fields of view at the same time. It can provide satellite attitude information and obtain satellite-geocentric vector information for navigation, etc. It is likely to replace the earth sensor, star sensor, etc., as the development direction of a new generation of sensors in the future and belongs to the forefront of research of optical sensors.

8.2   Satellite Independent Health Management

"Health status" describes the ability of the equipment and its subsystems and components to perform design functions. "Health status assessment" mainly refers to the comprehensive analysis of monitoring data and historical data obtained by various measurement methods, the evaluation of the health status of the equipment using various evaluation algorithms, and the unqualified reasons and operation suggestions for the unqualified equipment. By accurately understanding the health status of the current equipment, we can make a correct assessment of the health status of the equipment and provide a basis for the operation and decision-making of the equipment.

The satellite health assessment technology is very necessary and significant to ensure the highly reliable, autonomous, and stable operation of satellites. In the future, the single satellite function of the satellite navigation system will become increasingly complex, and the multi-satellite constellation system will appear in succession. If the satellite has a fault that affects the system service, the system service and availability may become poor, or even unusable, which will directly affect the application effect of the system. In addition, under the condition that the system has no ground support for a long time, if the satellite can assess the on-board health status, timely and effectively handle faults, and provide decision support for mission planning, maintenance, etc., it will effectively improve the viability of the system, which is of great value [7].

8.2.1 High Reliability and Autonomy of Satellites

Satellite is a complex system involving multi-disciplinary and multi-disciplinary technologies, and it is also flying in a harsh and complex space environment. Although a series of reliability measures have been taken in the design and development process, various failures still inevitably occur during the satellite's in-orbit flight, resulting in sudden or gradual deterioration of the satellite's health status. Statistics show that 156 of the 1036 satellites successfully launched from 1990 to 2009 have failed, accounting for 15% of the total number of satellites. The particularity of the space mission makes the failure of the whole mission possible if no measures are taken if the satellite fails during orbit. However, relying solely on the ground to take remedial measures, its effectiveness, and timeliness are limited, which may miss the best processing opportunity and lead to mission failure. If the health status assessment technology can be adopted on the satellite to find out the abnormal health status in time, and on this basis, the effective, and timely processing of faults is of great significance to the continuity, stability, and safety of the whole satellite system.

8.2.2 Long-term Autonomous Operation of Satellites in Orbit

At present, most of the on-orbit operation and management of satellites cannot be separated from the support of ground stations, relying on a large number of professional personnel to analyze and judge the telemetry data, and in case of failure, it also mainly depends on the decision of experts. However, under the constraint of not being supported by the ground system for a long time, as the basis of autonomous operation task execution, the satellite needs to have the ability of on-board health assessment to ensure a healthy and stable operation in orbit without ground support. Satellite health status assessment technology can analyze and predict the occurrence and development trend of faults, timely provide decision support for mission planning and maintenance in orbit, minimize and avoid the occurrence of serious faults, effectively reduce the impact of faults on the whole satellite, and effectively guarantee the long-term autonomous operation of satellites in orbit. In addition, it has great significance in reducing ground dependence and satellite operation costs [8].

The basic idea of health status assessment is to deeply integrate the collected health data information and low-level fault diagnosis/prediction results to obtain the health status assessment results. Health status assessment generally needs to be carried out according to the characteristics of the research object. Health status assessment methods mainly include model-based health status assessment method; analytic Hierarchy Process (AHP)-based health status assessment method, and AI-based health status assessment method.

1) Model-based health assessment method

   The model-based health status assessment method is a method of evaluating by establishing the physical or mathematical model of the studied object. The advantage of this method is that the evaluation results are highly reliable, but the modeling process is relatively complex, and the model verification is difficult. The model needs to be modified at any time as the evaluation object changes.

2) Health status assessment method based on AHP

   AHP is a quantitative decision-making method for the comprehensive evaluation of multiple indicators. It can express a complex problem as an orderly ladder hierarchy and quantify the qualitative factors by determining the initial weight of each evaluation index in the same level, which reduces the subjective impact to a certain extent and makes the evaluation more scientific.

3) Health assessment method based on artificial intelligence

   The health status assessment method based on artificial intelligence uses artificial intelligence methods to learn, reason, and make decisions based on data to achieve health status assessment, mainly including Bayesian network, fuzzy logic, artificial neural network, etc.

   (i) Health assessment method based on Bayesian network

   Bayesian network, also known as belief network, is an extension of Bayesian method and an uncertain knowledge representation model. The good knowledge representation framework of Bayesian network is the mainstream method of uncertain knowledge representation and reasoning technology in the field of artificial intelligence. It is considered to be the most effective theoretical model in the field of uncertain knowledge representation and reasoning at present. Bayesian network can easily deal with incomplete data problems, easy to learn causality, easy to achieve the integration of domain knowledge and data information, suitable for expressing and analyzing uncertain and probabilistic things. It can make inferences from incomplete or uncertain knowledge or information [9].

   (ii) Health status assessment method based on fuzzy logic

   When the evaluated object presents the fuzzy characteristic of "this is also that," the traditional accurate evaluation algorithm is difficult to apply, and the fuzzy evaluation method can solve this kind of problem very well. The general steps of the fuzzy evaluation method are: first, establish the factor set and reasonable evaluation set of the evaluation index, then obtain the fuzzy evaluation matrix through expert evaluation or other methods, and use the appropriate fuzzy operator to carry out the fuzzy transformation operation to obtain the final comprehensive evaluation result.

   (iii) Health assessment method based on artificial neural network

   By simulating the structure and function of biological neural system, artificial neural network has the characteristics of nonlinear, adaptive, fault-tolerant, parallel processing, and the ability to fully approximate any complex nonlinear relationship. It has not only the general computing ability to process numerical data, but also the thinking, learning, and memory ability to process knowledge. Because of its good nonlinear mapping ability, neural network has been widely used in pattern recognition, image processing, and other fields.

8.3 Intelligent Fault Diagnosis and Prediction Technology

The ultimate purpose of satellite health status assessment is to ensure the highly reliable and stable operation of the satellite. It is necessary to diagnose, isolate, switch or reconstruct the faults in time, predict the possible faults in the future, and make effective health status assessments accordingly for decision support. Therefore, the fault diagnosis method and fault prediction method are the basic technology to realize satellite health state assessment and also the key to realize health state assessment. Health status assessment is not limited to the whole satellite system level. The fault diagnosis and fault prediction results of subsystem level and key components are also important components of satellite health status assessment [10].

Fault diagnosis refers to the evaluation process of equipment operation status and fault status through comprehensive processing of various status information in the diagnosed equipment through certain methods. In essence, fault diagnosis is a process of pattern classification and recognition. In the past 40 years, fault diagnosis technology has rapidly developed into a new discipline. It is also an interdisciplinary discipline established to meet the needs of various projects, involving sensors and detection, signal analysis and data processing, artificial intelligence, automatic control, prediction, and other fields.

Fault diagnosis methods can be divided into three categories: fault diagnosis methods based on signal processing, fault diagnosis methods based on analytical model, and fault diagnosis methods based on knowledge.

8.3.1 Fault Diagnosis Method Based on Signal Processing

The fault diagnosis method based on signal processing uses the signal processing method to directly analyze the measurable signal of the object to be diagnosed. According to the directly measurable input and output and its change trend or the relationship between the directly measurable signal of the system and the fault, it is not necessary to establish an accurate mathematical analysis model. Its main idea is to collect and process the measurable signal of the object to be diagnosed, and analyze and judge based on the extracted features such as time domain and frequency domain to achieve fault diagnosis. The fault diagnosis method based on signal processing is the earliest and most widely used fault diagnosis algorithm for satellites.

In this kind of diagnosis method, the fault diagnosis method based on rule judgment is commonly used in practical engineering.

8.3.2 Fault Diagnosis Method Based on Analytical Model

The fault diagnosis method based on the analytical model is based on the mathematical analytical model of the system. Its basic idea is to establish a mathematical analytical model that can describe the system based on fully understanding and analyzing the principle, structure, function, and other relevant contents of the system to be diagnosed, and realize the fault diagnosis by comparing the actual output of the system with the model output. For example, the residual is generated by using observer (group), Kalman filter, parameter model estimation, and identification methods, and then the residual is evaluated or determined according to certain criteria or thresholds. However, it is often difficult to obtain accurate mathematical analytical models to describe complex systems.

The advantage of the model-based fault diagnosis method is that it can make full use of the deep knowledge in the system, which is conducive to accurately detect whether the fault occurs or not. The disadvantage of this method is that it relies too much on the mathematical model of the system and is sensitive to modeling error, parameter perturbation, noise, and interference. It is generally only applicable to linear objects and single fault analysis [11].

8.3.3 Knowledge-based Fault Diagnosis Method

Knowledge-based fault diagnosis methods make the diagnosis process knowledge-based. Under the premise of knowledge discovery, fault detection rules can be obtained by mining deep knowledge without establishing an accurate mathematical analytical model. They can be divided into two types: methods based on qualitative empirical knowledge and artificial intelligence methods based on quantitative knowledge. The former is represented by expert system. Its basic idea is to use the professional knowledge and experience of domain experts to establish corresponding rules and imitate the reasoning ability of fault behavior. The fault diagnosis method based on qualitative empirical knowledge has the bottleneck of knowledge acquisition, and the knowledge is difficult to maintain and lacks learning and adaptation mechanism. The latter uses artificial intelligence methods to learn, reason, and make decisions based on data to achieve fault diagnosis, so it is often called intelligent fault diagnosis. The commonly used methods mainly include artificial neural network, fuzzy logic, etc.

Fault prediction method is an important part of health state assessment. Prediction is to predict and speculate on things that have not yet happened or are not clear at present. It is to discuss and study the results that will happen in the system at present and give the development trend of the system's health state. In a sense, it is a kind of reasoning process. Fault prediction is a technical means to achieve fault prevention and improve the real-time performance of fault diagnosis. Fault prediction technology is mainly based on the analysis of the changing trend of key parameters of fatal or lossy faults to infer the possible state of satellites in the future, so as to take preventive measures to avoid the occurrence of satellite faults or reduce the harm of satellite faults.

Fault prediction methods can be divided into three categories: model-based fault prediction technology, data-driven fault prediction technology, and statistical reliability-based fault prediction technology.

8.3.3.1 Model-based Fault Prediction Technology

Model-based fault prediction refers to the prediction method using dynamic models or processes. Physical model method, Kalman/extended Kalman filter/particle filter, and expert-based method can all be classified as model-based fault prediction technology.

The model-based fault prediction technology generally requires that the mathematical model of the object system is known. This method provides a technical means to master the failure mode process of the predicted component or system, evaluate the loss degree of key components through the calculation of functional damage under the working conditions of the system, and evaluate the cumulative effect of failure in the use of components within the effective life cycle. By integrating physical model and stochastic process modeling, it can be used to evaluate the distribution of residual life of components.

Model-based fault prediction technology has the advantage of penetrating into the nature of the object system and realizing real-time fault prediction. In practical engineering applications, the mathematical model of the object system is often required to have high accuracy. However, the contradictory problem is that it is usually difficult to establish accurate mathematical models for complex dynamic systems. Therefore, the practical application and effect of model-based fault prediction technology are greatly limited, especially in the fault prediction of complex systems [12].

At present, model-based methods are mostly applied to electromechanical systems such as aircraft and rotating mechanisms. However, for complex systems, due to their relatively complex failure modes and failure mechanisms, the modeling research of fault prediction is relatively lagging.

8.3.3.2 Fault Prediction Technology Based on Data Driven

In many cases, the historical fault data or statistical data set caused by many different signals is the main means to master the system performance degradation. The method of prediction based on test or sensor data is called data-driven fault prediction technology. Typical data-driven fault prediction methods include regression analysis prediction method, gray model method, Markov method, support vector machine method, artificial neural network, fuzzy system, etc.

Unlike model-based methods, intelligent data-driven methods, such as neural networks, can achieve data adaptation. They can learn from samples and try to capture the intrinsic functional relationship between sample data and have achieved good application results in fault diagnosis and prediction.

The data-based fault prediction technology does not need prior knowledge of the object (mathematical model and expert experience). Based on the collected data, it uses various data analysis and processing methods to mine the hidden information for prediction operations, thus avoiding the shortcomings of the model-based and knowledge-based fault prediction technology. It becomes a more practical fault prediction method. However, typical data of some key equipment in practical applications often have strong uncertainty and incompleteness, which increases the difficulty of fault prediction technology.

8.3.3.3 Fault Prediction Technology Based on Statistical Reliability

The fault prediction method based on statistical reliability or probability is applicable to fault prediction from the perspective of statistical characteristics of past fault history data. Compared with model-based methods, this method requires less detailed information, because the information needed for prediction is contained in a series of different probability density functions, and does not need the form of dynamic differential equations. The advantage of this method is that the required probability density function can be obtained by analyzing the statistical data, and the obtained probability density function can provide enough support for the prediction. In addition, the prediction results given by this method contain confidence, which can also well represent the accuracy of the prediction results.

The typical failure probability curve based on statistical reliability is the famous "bathtub curve." That is, at the beginning of equipment or system operation, the failure rate is relatively high. After a period of stable operation, the failure rate can generally be maintained at a relatively low level, and then after a period of operation, the failure rate starts to increase again until all components or equipment fail or fail. The production characteristics of equipment, changes in historical tasks, performance degradation in the life cycle, and other factors make the fault prediction based on system characteristics more complex. All these factors will have a certain probability impact on the prediction results. In addition, it is also necessary to decrease and reduce the false alarm rate of fault prediction.

Fault prediction methods based on statistical reliability include Bayesian method, Dempster–Shafer theory, fuzzy logic, etc. All these methods are generally based on Bayesian theorem to estimate the probability density function of the fault. Through the reliability analysis of many engineering products and systems, the failure and time data trends of general products or systems are well subject to Weibull distribution. Therefore, Weibull model is widely used for the failure prediction of systems or equipment.

8.4 Intelligent On-Orbit Maintenance of Satellite

8.4.1 Evolutionary Hardware Technology

The use of intelligent evolutionary algorithms in hardware circuits can make them adaptive and fault-tolerant. The technology of applying intelligent evolutionary algorithms to the construction of circuit systems in reconfigurable devices is called "Evolvable Hardware" (EHW).

The idea of evolutionary hardware originated from the idea of developing a computer with the ability of self-reproduction and self-repair put forward by Von Neumann, the father of computer in the 1950s, which was not realized due to the technical conditions at that time. With the emergence of evolutionary algorithms and large-scale programmable logic devices, the idea gradually became possible to be realized. By the 1990s, De Hugo, the father of artificial brain, and scientists from the Swiss Federal Institute of Technology, proposed the concept of evolutionary hardware. Evolutionary hardware refers to a hardware system that can automatically change its structure and function according to the current environment to adapt to environmental changes. It is mainly composed of reconfigurable hardware and configuration engine, as shown in Figure 8.2. As a configuration engine, evolutionary algorithm can dynamically change the hardware structure according to the performance of the hardware system and then change the hardware functions to adapt to the current environment changes [13].

Since the concept of evolutionary hardware was put forward, there have been two views on its development direction: first, the goal of evolutionary hardware is to find and design circuit structures superior to traditional methods, so as to realize the automation of current large-scale integrated circuit design and get rid of the current situation that circuit design depends on technicians. The second is that the goal of evolutionary hardware is to design online systems with self-healing and adaptive capabilities to meet the needs of deep-sea, aviation, and other fields. Therefore, the research of evolutionary hardware is based on two methods: Darwin's theory of evolution and cell regeneration theory. The former research institutions are mainly the Japanese Institute of Electronic Technology,

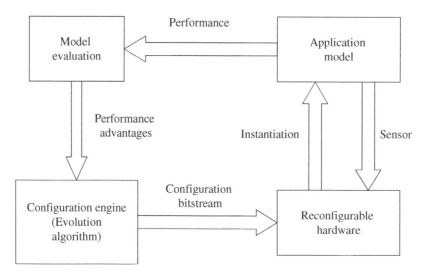

Figure 8.2 Structure of evolutionary hardware.

the American Jet Engine Propulsion Laboratory, the University of Sussex in the United Kingdom, and the University of Heidelberg in Germany. The latter research institutions are mainly the University of York in the United Kingdom and the Swiss Federal Institute of Technology.

Evolutionary hardware is different from the hardware that executes evolutionary algorithms. The hardware structure of the latter will not change. It is mainly used to perform evolutionary operations, such as selection, crossover, and mutation in genetic algorithms, to speed up the speed of evolutionary algorithms. Basically, evolutionary hardware is also different from using evolutionary algorithms to optimize parameters. The goal of evolutionary hardware is to realize the adaptation of the hardware system to meet the needs of the environment [14].

Evolutionary hardware mainly includes evolutionary algorithms and reconfigurable hardware. As the configuration engine of evolutionary hardware, evolutionary algorithms currently used for evolutionary hardware mainly include genetic algorithms (GA), genetic programming (GP), evolutionary programming (EP), and evolutionary strategies (ES). In terms of reconfigurable hardware, according to different applications, there are mainly commercial platforms such as FPGA, FPAA, and FPTA, as well as application platforms specially developed for evolutionary hardware.

Evolutionary hardware can dynamically change its structure to adapt to environmental changes. Therefore, evolutionary hardware uses evolutionary algorithms and reconfigurable hardware. Reconfigurable hardware configures its own

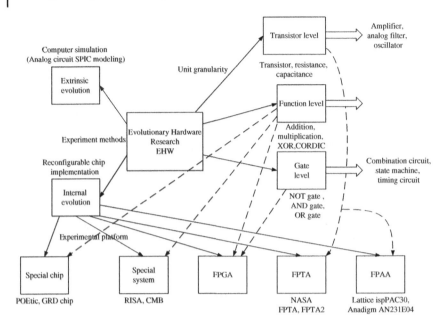

Figure 8.3 Classification of evolutionary hardware research.

structure according to the structure bit string corresponding to its own structure. Different structure bit strings correspond to different hardware structures. Under the control of evolutionary algorithm, the structure of bit strings is changed to obtain different hardware structures, to obtain a hardware system that meets the requirements (Figure 8.3).

The basic idea of evolutionary hardware is to take the structural bit string representing the topological structure and attributes of hardware as the basic object of evolutionary algorithm, such as the chromosome of genetic algorithm, and evaluate the performance of the hardware structure represented by the chromosome by software simulation or hardware measurement, to guide the evolution process with the fitness function. The evolutionary hardware model is shown in Figure 8.4.

The reconfigurable hardware depends on the application situation, and the platform selected is also different. FPGA is used for digital circuit design, FPAA is used for analog circuit design, and FPTA is used for hybrid circuit design. Although the hardware platforms vary greatly, the basic structure is roughly the same. Figure 8.5 is a simplified FPGA hardware structure diagram, which contains a set of two-dimensional hardware reconfigurable units (FBs) and wiring resources around each unit. The configuration switch (small black dot on the wire) determines the connection mode of the input and output of the reconfigurable unit. There are also

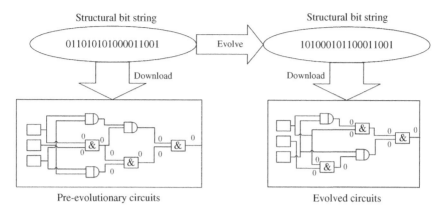

Figure 8.4 Evolutionary hardware model.

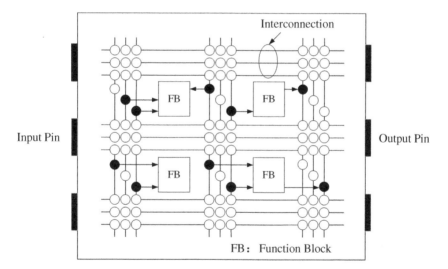

Figure 8.5 Simplified FPGA structure.

configurable function selection switches in each reconfigurable unit, each of which represents a functional module, such as AND gate, OR gate, and NOT gate. The specific connection mode and function selection configuration are stored in a peripheral storage block, which is controlled by software reading and writing. By reading and writing the storage area, the corresponding hardware circuit structure can be realized. The evolutionary algorithm controls the circuit structure of the hardware by reading and writing the corresponding region. The performance evaluation of circuit structure can be divided into external evolution and internal evolution according to whether it is evaluated on the actual circuit. External evolution calculates the fitness value of the corresponding circuit structure through software simulation. Internal evolution downloads the corresponding circuit structure to the hardware platform and obtains its fitness value through the actual evaluation results. The external evolution is relatively safe and does not stick to the specific hardware structure. The internal evolution examines the hardware performance, so the results are accurate and fast, but it may cause damage to the hardware system (Figure 8.5).

With the development of evolutionary algorithms, more algorithms are selected for hardware evolution, such as GA, GP, ES, and EP. GA and GP emphasize the changes and interactions of chromosomes in evolution, ES focuses on behavioral changes at the individual level, while EP emphasizes behavioral changes at the population level. In terms of algorithms, their differences mainly lie in individual construction and evolutionary operation. In recent years, some new algorithms, such as particle swarm optimization (PSO), Hereboy, and gene expression programming (GEP), have also been used in evolutionary hardware. GA have been popular in the industry since the concept of hardware evolution was proposed. GP has unique advantages in the design of complex structures such as analog circuits, but its operation is complex, and its evolution time is long. In the two hot areas of evolutionary hardware research, namely, system design and online adaptation and fault tolerance, the algorithm requirements are also different due to its own characteristics. The main goal of system design is to obtain optimal design results, with no special requirements for the running time and complexity of the algorithm. However, in terms of system adaptation and fault tolerance, especially in the application of embedded systems, not only the efficiency and speed of the algorithm should be considered, the algorithm is also required to have a small implementation cost.

8.4.2 Reconfigurable Computing Technology

With the continuous development of aerospace technology and information technology, satellite functions are increasingly complex and diversified. In addition to the traditional features of high reliability and long life, flexible, efficient, and rapid response have become the application requirements and development direction of satellites. Reconfigurable computing technology can realize the flexible and variable hardware system in orbit and effectively improve the complex functional adaptability of satellites.

The concept of reconfigurable computing, which is widely accepted by the academic community, was put forward by André Dehon and John Wawrzynek of the University of California, Berkeley, in 1999. This definition is given by comparing with traditional computing models: (i) It is different from ASIC: reconfigurable computing can be customized for computing tasks after hardware manufacturing. (ii) Different from general-purpose processors: reconfigurable computing can provide a lot of customization space for the mapping of algorithms to hardware. In 2002, Ketherine Compton and Scott Hauck gave a more specific definition: use the system integrated with programmable hardware to calculate, and the function of the programmable hardware can be defined by a series of physical controllable points with timing changes.

The idea of reconfigurable computing was first proposed by Gerald Estrin of the University of California, Los Angeles, in 1962. However, due to the lack of support for reconfigurable device technology, the reconfigurable system it implements is only an approximation of the design concept of reconfigurable computing. In the mid-1980s, reconfigurable computing technology gradually gained wide application after Xilinx Company introduced the field programmable gate array technology (FPGA) and Altera Company introduced the complex programmable logic device (CPLD). With the further maturity of FPGA technology, FPGA has become the mainstream hardware platform of reconfigurable computing.

Reconfigurable computing did not get enough attention at the beginning of its emergence. Until 1992, the Splash 2 system designed by the American Supercomputer Center based on Xilinx's FPGA was 2500 times faster than the SPARC 10 workstation at that time in the application of genome analysis calculation and gray image median filter. This amazing result has aroused the interest of academia and industry in reconfigurable computing. With the continuous development of information technology, reconfigurable computing has been widely applied to the accelerated computing of various computationally intensive algorithms such as genome matching, high-energy physics, image processing, financial data processing, cloud computing, machine learning, etc. Its flexible and efficient characteristics have been concerned by many fields.

The classic book Reconfigurable Computing: The Theory and Practice of FPGA-Based Computing in the field of reconfigurable computing summarizes reconfigurable computing into three research directions: reconfigurable computing hardware, reconfigurable computing system programming, and application research. Each research direction also contains several research topics.

8.4.2.1 Research on Reconfigurable Computing Hardware
The research of reconfigurable computing hardware includes the research of reconfigurable devices and reconfigurable computing systems.

Reconfigurable devices are the basis of building reconfigurable computing systems. The reconfigurable cell structure, interconnection, and configuration methods inside the reconfigurable devices have a significant impact on the performance of reconfigurable computing systems. In addition to the widely used

FPGA structure, scholars have also studied customized reconfigurable logic devices with different characteristics, such as Garp, RapiD, PipeRench, etc. However, considering the economy and the support of development tools, FPGA is the main hardware platform for reconfigurable computing.

Reconfigurable computing system refers to a computing system composed of single or multiple reconfigurable devices, which is the hardware basis for completing specific computing tasks. The computing system composed of multiple reconfigurable devices is currently developing toward supercomputing. This type of computing system is generally composed of multiple FPGAs and high-performance general-purpose processors. The interconnection mode, data storage mode, memory bandwidth, and control bus bandwidth with the master controller of multiple FPGAs are key issues to be solved. At present, some commercial products have appeared, such as the XD1 of Cray Company and the BEE3 system of the University of California, Berkeley, etc.

With the isomerization development of the internal structure of FPGA, Block RAM, PowerPC processor, etc., are embedded into the FPGA. The reconfigurable computing system composed of a single FPGA is more in line with the requirements of high-performance computing under the strict restrictions of volume, power consumption, and weight in the embedded environment. The design method, architecture, configuration, and implementation method of monolithic reconfigurable computing systems have not yet been solved by common standards, which is a hotspot in the field of reconfigurable computing.

8.4.2.2 Research on Reconfigurable Computing System Programming

The research of reconfigurable computing system programming is to carry out research on the development tools, development environment, and hardware resource management of reconfigurable computing systems after completing the design of reconfigurable computing system. It is the software foundation for reconfigurable computing systems to realize computing functions and provide system performance. The research of reconfigurable computing system programming includes programming language research, high-level language compilation, layout and routing algorithm and task scheduling research. The first three are aimed at the design input, compilation, and hardware mapping stages in the development process of reconfigurable computing system. The main purpose of the research is to solve the problems of high design complexity and low development efficiency in the development methods based on traditional hardware description languages (such as Verilog HDL, VHDL, etc.). At present, some research results have been published, such as SystemC, HandleC, and AutoESL, but the research in this field is not yet mature, and there are still problems such as high hardware occupancy of compilation results, low execution efficiency, and no support for dynamic layout and routing.

Task scheduling is to complete hardware and computing task management at the reconfigurable computing system level. Effective task scheduling is the key to improving hardware utilization and task computing efficiency. The optimal task scheduling of a computing system is an NP-complete problem. Due to the characteristics of its space-time computing mode, reconfigurable computing systems need to achieve scheduling in both time and space dimensions. In addition, reconfigurable computing systems still have resource constraints, configuration delays, and the number of configured ports, which make their task scheduling problems more complex.

8.4.2.3 Applied Research

Application research mainly includes application research oriented to specific fields and comparative research with other computing modes. In the application of specific fields, researchers are generally concerned about how to use limited computing resources to improve computing efficiency for specific application problems. This research involves algorithm structure design, calculation methods of nonlinear functions, and adaptive improvement of algorithm to the FPGA platform. At the same time, to compare the effectiveness of different computing modes in different applications, another topic of research on reconfigurable computing applications is to carry out comparative research on computing efficiency, power consumption, and resource utilization between reconfigurable computing and other computing modes, such as general processor platform and emerging GPU platform.

According to the time of reconfigurable hardware reconfiguration (configuration) in reconfigurable system, reconfigurable computing system can be divided into static reconfiguration mode and dynamic reconfiguration mode. Different reconfiguration modes are suitable for different application scenarios. At the same time, the development of reconfigurable device reconstruction technology also provides support for the above two reconstruction modes.

Reconfigurable Mode of Reconfigurable System Static refactoring, also known as compile time reconstruction. The configuration information must be configured to the logic resources of the reconfigurable logic device before the device starts to perform the operation. During the whole operation process of the device, the configuration on the reconfigurable logic device will remain static without any change. Static reconfiguration systems generally have no strict requirements for configuration time.

Dynamic reconfiguration, also known as runtime reconfiguration, is characterized by the ability to reconstruct logical resources on reconfigurable logic devices while tasks are running. The dynamic reconfiguration system can be divided into global dynamic reconfiguration system and partial dynamic reconfiguration system according to whether to reconstruct all on-chip logical resources. With the dynamic reconfiguration method, the task can be divided into multiple units, and each unit forms a separate configuration file. According to the execution requirements of the task, the configuration files representing different task units are loaded into the reconfigurable devices respectively, to realize the "big task" with "small hardware." However, to ensure the continuity of task execution, the reconfiguration time of dynamic reconfiguration system is generally short.

Compared with static reconfiguration technology, dynamic reconfiguration technology is more flexible and efficient, and can improve the utilization of hardware. It is the mainstream direction of the development of reconfigurable computing technology.

Reconfiguration Technology of Reconfigurable Devices   The reconfiguration technology of reconfigurable hardware can be divided into the following categories:

1) **Single context**: Single context FPGA uses serial bit stream to configure programmable bits on the device, and any change in the configuration of programmable bits on the FPGA requires reconfiguration of the whole FPGA. Many shelf FPGA products are single-context devices, such as Xilinx 4000 series, Altera Flex10K series, and Cyclong series. Because the single context device needs to reconfigure the whole device every time, its configuration time cost is high, and it is generally used in the static reconfiguration mode, and of course, it can also be used in the dynamic reconfiguration system that does not require strict reconstruction time.

2) **Multi-context**: Multi-context FPGA provides multiple memory of configuration information for each programmable bit. Different configuration information memory can be understood as different storage layers of configuration information. The configuration information of a layer can become active at a given time, and the FPGA device can switch the configuration information of different layers in a short time. From the perspective of working principle, the modification of each configuration bit by the multi-context FPGA also requires the reconfiguration of the whole device, but the reconfiguration process is fast and the configuration time is less, so the multi-context device can be applied to the global dynamic reconfiguration system. However, FPGA devices need to design more configuration information storage bits, which limits the logical resource scale of the device. At present, there have been some research results that support multi-context, such as Morphosys, Paddi, Cameron, and Rapid, but most of the above results are in the experimental stage, and there is still a lack of practical commercial products.

3) **Partially reconfigurable**: In partially reconfigurable FPGAs, the programming bits generally use SDRAM technology, which can directly address the

configuration data, so that specific programming bits can be reconfigured without the need to reconfigure the entire FPGA. Because the partial configuration process of the partially reconfigurable FPGA only affects the reconfigured hardware resources, while other hardware resources can still work normally according to the original configuration, the partial reconfiguration allows the work to be carried out simultaneously with the reconfiguration, thus supporting the dynamic reconfiguration system. On the other hand, because part of the reconfiguration process only loads part of the configuration file, that is, only changes the function of part of the logical resources of the FPGA, it can support part of the dynamic reconfiguration system. Xilinx, a major FPGA manufacturer, started with its Virtex II Pro series, and all subsequent series of FPGAs have supported partial reconstruction technology.

Using the reconfigurable hardware that supports partial reconfiguration to build a partially dynamic reconfigurable computing system, in addition to the flexibility and high efficiency of reconfigurable computing systems, it can also achieve large-scale applications on small-scale devices, thus improving the hardware utilization. At the same time, the characteristics of partial reconfiguration enable the partial dynamic reconfiguration system to support real-time multitasking. In view of the advantages of some dynamic reconfigurable systems in terms of flexibility, computing efficiency, hardware utilization, and real-time multitask support, it has become the mainstream development direction of reconfigurable computing, and has been widely supported by commercial FPGA devices and development tools.

8.5 Typical Application Cases

8.5.1 Long-term Autonomous Navigation

Autonomous navigation refers to the ability and process of ensuring the autonomous and stable operation of the navigation system by using the preannotated ground auxiliary data, by means of inter-satellite/satellite-ground measurement, inter-satellite/satellite-ground data exchange, and satellite autonomous navigation software processing under the condition that the satellite cannot receive the normal support of the ground system for a long time. Compared with general software, spaceborne autonomous navigation software has complex features such as continuous operation for a long time, multiple task conditions, dynamic functional flow, large amount of data interaction, and strict timing constraints. With the full completion of the Beidou-3 system, the autonomous navigation function of the constellation will carry out the operation of 30 stars in the whole network. In this process, the processing of numerous complex data and the resulting system refinement need to use big data processing methods and some artificial intelligence algorithms [15].

In the next-generation system demonstration, the centralized autonomous navigation supported by the laser inter-satellite link will achieve higher system integration, more comprehensive and reasonable information allocation, more sophisticated accuracy, higher robustness, and higher intelligence. AI algorithms and technologies will be highly integrated with the above functions and indicators, such as inter-satellite network topology optimization and autonomous planning technology, massive inter-satellite data processing kick field reduction technology, centralized task processing, and seamless task handover [16].

8.5.2 Navigation Intelligent Countermeasure

The navigation intelligent countermeasure technology can detect, locate and recognize the jamming signals based on the navigation satellite's intelligent perception ability of malicious jamming signals, quickly and flexibly adjust the satellite navigation signal spectrum, power and shape, and form intelligent linkage feedback with ground users, effectively improving the countermeasure ability in complex battlefield environment [17, 18].

References

1 Garrison, W.G. (2008). *Attaining Fault Tolerance Through Self-adaption: The Strengths and Weaknesses of Evolvable Hardware Approaches*, Part of the Lecture Notes in Computer Science Book Series, vol. 5050, 368–387.

2 Antonio, M. (2008). Introduction to evolvable hardware: a practical guide for designing self-adaptive systems. *Genetic Programming and Evolvable Machines* 9: 275–277.

3 Jason, L., Greg, L., and Ronald, D. (2003). A genetic representation for evolutionary fault recovery in Virtex FPGAs. In: *ICES 2003, LNCS 2606*, 47–56.

4 Estrin, G., Bussel, B., Turn, R. et al. (1963). Parallel processing in a restructurable computer system. *IEEE Transactions on Electronic Computers* 12 (5): 747–755.

5 Lynn, A.A., Athanas Peter, M., Luna, C. et al. (1994). Finding lines and building pyramids with Splash2. In: *Proceedings of the IEEE Workshop on FPGAs for Custom Computing Machines*, 155–163. IEEE.

6 Alachiotis, N., Berger, S.A., and Stamatakis, A. (2011). Accelerating phylogeny-aware short DNA read alignment with FPGAs. In: *Proceedings of the IEEE International Symposium on Field-Programmable Custom Computing Machines (FCCM)*, 226–233. IEEE.

7 Kirsch, S., Rettig, F., Hutter, D. et al. (2010). An FPGA-based high-speed, low-latency processing system for high-energy physics. In: *Proceedings of the International Conference on Field Programmable Logic and Applications (FPL)*, 562–567. IEEE.

8 Haohuan, F. and Clapp, R.G. (2011). Eliminating the memory bottleneck: an FPGA-based solution for 3D reverse time migration. In: *Proceedings of the 19th ACM/SIGDA International Symposium on Field Programmable Gate Arrays (FPGA'11)*, 65–74.

9 Pottathuparambil, R., Coyne, J., Allred, J. et al. (2011). Low-latency FPGA based financial data feed handler. In: *Proceedings of the IEEE 19th Annual International Symposium on Field-Programmable Custom Computing Machines (FCCM)*, 93–96. IEEE.

10 Madhavapeddy, A. and Singh, S. (2011). Reconfigurable data processing for clouds. In: *Proceedings of the IEEE 19th Annual International Symposium on Field-Programmable Custom Computing Machines (FCCM)*, 141–145. IEEE.

11 Orlowska-Kowalska, T. and Kaminski, M. (2011). FPGA implementation of the multilayer neural network for the speed estimation of the two-mass drive system. *IEEE Transactions on Industrial Informatics* 7 (3): 436–445.

12 Hauck, S. and DeHon, A. (2008). *Reconfigurable Computing: the Theory and Practice of FPGA-Based Computation*, 475–496. San Fransisco: Morgan Kaufmann.

13 Callahan, T.J., Hauser, J.R., and Wawrzynek, J. (2000). The Garp architecture and C compiler. *Computer* 33 (4): 62–69.

14 Ebeling, C., Cronquist, D.C., and Franklin, P. (1996). RaPiD: Reconfigurable pipelined datapath. In: *Proceedings of the 6th International Workshop on Field-Programmable Logic and Applications*, 126–135.

15 Goldstein, S., Schmit, H., Moe, M. et al. (1999). PipeRench: a coprocessor for streaming multimedia acceleration. In: *Proceedings of the 26th International Symposium on Computer Architecture*, 28–39.

16 Davis, J.D., Thacker, C.P., and Chang, C. BEE3: Revitalizing Computer Architecture Research. https://www.microsoft.com/en-us/research/publication/bee3-revitalizing-computer-architecture-research/ (accessed April 2009).

17 Liu, C.L. and Layland, J. (1973). Scheduling algorithm for multiprogramming in a hard-real-time environment. *Journal of the ACM* 20 (l): 46–61.

18 Papadimitriou, K., Dollas, A., and Hauck, S. (2011). Performance of partial reconfiguration in FPGA systems: a survey and a cost model. *ACM Transactions on Reconfigurable Technology and Systems* 4 (4): 1–28.

9

Application of AI in Aerospace Loads

9.1 Intelligent Load Software Architecture

The software architecture of the intelligent payload satellite system supporting APP is the core of the intelligent payload satellite system to realize intelligence. The intelligent payload satellite system needs the support of three basic software parts to provide services for users [1].

1) Running on the user client APP (APP-USER, user terminal software), it can support users to send user requirements to the intelligent satellite system through the terminal and can also receive information feedback from the intelligent satellite system and display it to users.
2) The onboard APP (APP-SAT, onboard application software) running on the satellite can complete the in-depth processing of image data according to user needs to form images or data of user concern.
3) The server-side APP (APP SERVER, server software) can cooperate with the satellite APP to complete further image data processing and then send it to the user.

The effective operation of the above three parts of the APP needs the support of the onboard and ground software architecture [2].

The intelligent payload satellite software is divided into two parts: the intelligent management software running in the satellite platform management unit and the intelligent processing software running in the onboard intelligent processing platform. The satellite platform management software is mainly responsible for the routine management of the satellite platform. At the same time, it completes the autonomous task planning and execution according to the mission instructions of the cloud service center and provides the imaging results to the onboard intelligent processing platform. Satellite platform management software

Intelligent Satellite Design and Implementation, First Edition. Jianjun Zhang and Jing Li.
© 2024 The Institute of Electrical and Electronics Engineers, Inc. Published 2024 by John Wiley & Sons, Inc.

is mainly composed of embedded real-time operating system, satellite service management software, autonomous fault handling software, onboard autonomous mission planning to software, etc.

The onboard intelligent processing software is mainly composed of embedded operating system, payload data service software, platform information service software, onboard APP scheduling management software, and onboard APP, as shown in Figure 5.2. The core of the software framework is the onboard APP scheduling management software, whose functions include: receiving and installing a series of APP software packages from the cloud service center through the satellite-ground measurement and control channel. Manage the running and stopping of APP software, and delete APP software. Transfer load data and platform information (platform time, orbit, attitude, and other information) to APP software. The processed data received from APP software is transmitted to the ground through a satellite data transmission channel [3].

In order to expand the application of intelligent payload satellite and provide users with more extensive services, the APP of intelligent payload satellite needs to learn from the operation mode of APP STORE, that is, to build an open platform. First, it is convenient for third-party developers to develop APP based on this platform to obtain the resources of payload satellite or real-time processing of payload satellite resources in orbit. Second, clarify the profit model of third-party developers, so that developers can obtain benefits from it while providing services for users, maximize the enthusiasm of third-party developers, deeply explore the services that payload satellites can provide for the general public, and form a benign ecological environment for development, so as to continuously develop a variety of APP.

9.2 Cloud Service Center Software Architecture

The cloud service center is an important part of the intelligent satellite system. The cloud service center is built by the intelligent satellite system operator. The software architecture is mainly composed of the operation server operating system, intelligent satellite system management and maintenance software, onboard APP management platform, server-side APP management platform, client-side APP management platform, user task request receiver service software, user task decomposition and satellite task scheduling software, satellite historical data storage and retrieval service software [4].

The cloud service center software mainly has the following functions:

1) Connect with satellite ground TT&C station and data transmission station through ground network to manage and maintain the entire intelligent payload satellite (constellation) system.

2) It can simultaneously receive the concurrent requirements of multiple user terminal software (APP-USER) through the network, complete the task decomposition through the satellite state estimation and intelligent configuration of resources, and send the decomposed user tasks to the intelligent payload satellite through the satellite-ground link. In this process, users do not need to pay attention to the operation status of the satellite system, nor do they need to pay attention to whether the satellite is serving other users.

3) Receive the image or data processed by the satellite onboard APP, and jointly complete the further processing of the image or data with the service software server-side APP running by the cloud service center, and then send the processed image or data to the user terminal software client APP through the network.

4) The image or data received from the satellite is stored for a long time, and the user's request data is appropriately cached. The long-term stored data can be used by users who need long-term data analysis. The data cache can provide fast data feedback for users who have the same request and reduce the direct data request to the satellite.

In addition, the cloud service center also needs to manage and maintain the APP, including receiving the APP software developed by third-party developers based on user needs through the network, including server-side APP, client-side APP, and satellite-borne APP, which are stored in the cloud service center. For client APP, the cloud service center provides software APPSTORE functions, namely storage and maintenance, provides download services for users, and provides upload and update services for APP developers. For the onboard APP, the cloud service center provides the verification and upload function, that is, after the security and reliability verification of the software is completed, the onboard APP is injected into the intelligent payload satellite through the network and the satellite-ground link. For the server-side APP, the cloud service center provides the operating environment, that is, the software can run on the hardware platform of the cloud service center, and cooperate with the onboard APP and the client APP to complete the image or data processing [5].

In a broad sense, it is generally believed that using computers to simulate human intelligent behavior belongs to the category of artificial intelligence. During the observation process of the load system, there are various unexpected tasks and observation environments and objects that change at any time. These situations are analyzed and considered by people one by one, and the workload is large and difficult to complete. Through artificial intelligence, the autonomous decision-making processing of the load work on orbit can save time, ensure the completion quality of the observation task, reduce the number of personnel and equipment, reduce the system overhead, and effectively improve the work efficiency of the system [6].

The intelligent payload satellite and cloud service center are provided and maintained by the intelligent satellite system operator, and third-party APP developers can develop a variety of APP software to facilitate users to obtain different services. The system operation scenario is designed as follows:

1) The user sends the user's business requirements (including imaging area, processing requirements, etc.) to the ground cloud service center through the mobile terminal APP. The cloud service center makes preliminary planning according to the resources of the intelligent payload satellite system (ground system, constellation system, adjacent satellites, time planning), etc., gives a list of satellites expected to complete the imaging task, and sends the task to the satellites in the list.
2) The intelligent payload satellite that receives the task completes the autonomous mission planning according to its own orbit, illumination and other conditions, and combines the target, and feeds back the planning situation to the ground cloud service center. After the ground cloud service center confirms, it selects the best execution satellite and sends the execution instruction to the satellite. After receiving the execution instruction, the intelligent payload satellite starts to complete the observation task according to the instructions of the autonomous planning, and obtains the image or video information.
3) On the basis of acquiring images or videos, the intelligent satellite calls the corresponding onboard APP processing software according to the processing requirements of the user's APP, completes the intelligent processing of the payload data, forms the user's focused intelligence information, and transmits the processed results to the ground cloud service center through the link.
4) The ground cloud service center transmits the user's final demand information to the user's mobile terminal through the ground network and wireless network.
5) If the processing capacity of the data acquired by the intelligent payload satellite is limited, it can be sent to the ground cloud service center for high-speed processing, and then the ground center will send the information to the user.

In this envisioned future scenario, the final state of the system operation is that the user (personal, commercial or military application) and the task demand and feedback process of the intelligent load system are completed in a short time, and the entire system operates intelligently, striving to be unattended.

9.3 Network-oriented Communication Protocol

The Consultative Committee on Space Data System (CCSDS) has carried out research on network communication protocols for satellite internet access to adapt to space applications. The 2012 CCSDS recommended IP Over CCSDS space link standard, which proposed the encapsulation of IP packets. Then, through the existing underlying link transmission idea of CCSDS, by providing relevant services and protocols, the problem of interconnection of heterogeneous networks in space tasks is solved, and the networked interaction between space communication entities is realized, which can provide end-to-end data transmission services [7].

For the network access of intelligent payload satellites, the satellite-ground link can adopt the CCSDS protocol architecture. The data link layer can choose the remote-control spatial data link layer protocol (TC SDLP), the telemetry spatial data link layer protocol (TM SDLP) or the advanced on-orbit system spatial data link layer protocol (AOS SDLP). The network layer adopts the IP protocol encapsulated by the encapsulated service and IP Over CCSDS. The transmission layer selects the space communication protocol specification – transmission protocol (SCPS-TP) Transmission Control Protocol (TCP) or User Datagram Protocol (UDP) can be customized by the application layer according to the actual application. After adopting the IP protocol encapsulated by IP Over CCSDS, the intelligent payload satellite can join the ground network through the satellite-ground link and become a member of the network. It can assign an IP address to the payload satellite. The ground can access the intelligent payload satellite through the IP address, and can even browse the real-time data or stored file data of the satellite like a web page. Ethernet communication is adopted between the ground station and the cloud service center. The cloud service center can analyze the application layer protocol. It can analyze and process the data from the payload satellite APP, obtain image information, and upload the APP software that needs to be injected on the satellite to the intelligent payload satellite through the ground station and the satellite-ground link. The terminal user can access the ground Ethernet through the mobile phone network or WIFI and other wireless networks to achieve data interaction with the cloud service center, receive the smart payload satellite APP data, or send the task request of the terminal APP to the smart payload satellite [8].

9.4 Intelligent Expert System

The satellite load intelligent system includes three parts: intelligent expert system, intelligent execution system, and intelligent semantic interpretation system, including full-field sensing instrument, zoom medium/high-resolution imager, and variable spectral resolution imager. The intelligent semantic interpretation system is mainly used for the barrier-free communication of users [9].

The intelligent expert system is the brain of the load system. Through the intelligent recognition module, the data obtained from the predetection of the full-field sensor is analyzed on the orbit, the geometric features in the image are compared with the features in the feature database, and the target features are extracted and recognized to judge the target attributes. The intelligent decision module determines the value of the target and gives the best imaging parameters of the payload, which is used to command the intelligent execution system. The intelligent execution system receives the instructions of the intelligent expert system, drives the zoom medium/high-resolution imager and the variable spectral resolution imager to track and detect the specified target. At the same time, the work of the intelligent expert system needs to carry out self-evaluation and ground-up evaluation. Through the analysis of the historical evaluation data, the experience of "numerical" is summarized to guide the next expert system work, that is, the intelligent evolution module [10].

9.4.1 Intelligent Identification Module

The function of the intelligent recognition module is to automatically analyze the image information by means of computer information processing, so as to find the object of interest and confirm the object type. The intelligent recognition module is divided into three parts: image target area capture, image feature extraction and image feature recognition. The target area capture link extracts the valuable areas in the remote sensing image from many useless information. The image feature extraction link extracts the target features from the valuable target area information. The image recognition module classifies and describes the image according to the geometric and texture characteristics of the figure using recognition theories such as pattern matching and discriminant function.

9.4.2 Intelligent Decision-Making Module

Which working mode should be the best for a certain target and whether the parameter setting is the best are the system problems for the evaluation of target value and the planning of stress plan. In an image, because there may be multiple targets at the same time, some of them are more concerned by users, such as airports and ports, while others can be ignored. The importance of these objectives can be quantitatively expressed by the objective preference function. First, input the output of the target capture phase, namely the recognition vector, into the target preference function, and the system will determine which targets can be ignored and which target needs to be tracked. Then the system extracts the important objects and retrieves the cases in the case database to detect the similarity. According to the modification rules, the historical decision is modified with reference to the current environment, and then the quasi-decision vector is obtained [11].

The decision indicators are extracted from the pseudo-decision vector, and the corresponding detection mode, direction, and spectral segment combination planning scheme is estimated by a specific algorithm to determine whether it meets the target threshold. If it does not meet the target threshold, the above process is repeated, and the mode planning is repeated until the target threshold is met, and the next step of the parameter and path stress planning system is entered.

The parameter and path stress planning system include the establishment of a change algorithm, obtaining the function based on the parameter and path preference vector, and introducing the stress imaging simulation vector based on the parameter and path, and obtaining the optimal parameter and path combination through the specific algorithm cycle parameter and path planning. The best mode, parameter and path planning scheme are used as the final onboard stress decision vector, enter the intelligent execution system, and return to the correction case database [12].

9.4.3   Intelligent Evolution Module

One of the main characteristics of intelligent systems is that they can adapt to unknown environments, and learning ability is one of the key technologies of intelligent systems. The intelligent evolution system on the satellite needs to have the ability of self-judgment, self-learning, and self-renewal. Among them, self-learning, and self-renewal are the feedback based on image information obtained by the intelligent system after deciding on the satellite and taking a stress response to independently judge the effect of a certain decision and stress response. If it reaches the default threshold of the system, that is, the decision and response are correct, then the decision and stress response will be updated into the strategy database for the next visit. This intelligent evolution scheme is based on reinforcement learning technology, and is a special learning method that takes environmental feedback as input and adapts to the environment. For reinforcement learning, its goal is to learn a stress strategy in a new environment [13].

9.5   Intelligent Execution System

Intelligent functions such as target recognition, automatic tracking, and detailed investigation of the satellite payload system based on artificial intelligence will significantly improve the efficiency and timeliness of the satellite and meet the user's needs to the maximum extent. The intelligent execution system receives and executes the instructions of the intelligent expert system to complete the imaging task. The intelligent execution system includes large-area low-resolution imaging, local medium-resolution imaging and high-resolution imaging of key targets. Among them, low-resolution imaging realizes wide area perception and automatically locates the target area. Medium-resolution imaging realizes the medium-resolution search of the target area and the positioning of key targets. High-resolution imaging realizes high spectral resolution recognition of key targets. Taking river basin observation as an example, the working process of intelligent satellite load system is introduced below [14].

The load system receives the user's instructions, and the intelligent execution system drives the full-field sensing instrument to perform large-area low-resolution imaging. The intelligent expert system receives the image data of the intelligent execution system and analyzes the data in orbit through the intelligent recognition module. The specific method is to divide the remote sensing image into regions based on the uniform block algorithm, gradually eliminate other information through the feature extraction algorithm, and "find" curve features in each region. According to the judgment criteria of the river curve, the target of "locking" is the river. The intelligent decision module evaluates the value of the target according to the target feature vector and enters the onboard decision system. Through the vector coordinate inversion of the target in the image, the direction change information of the optical system is obtained. According to the low-resolution image information, the exposure, resolution and detection mode are analyzed, and finally, the onboard decision is formed, and the instructions are sent to the intelligent execution system to obtain the high-resolution image of the valuable target. The intelligent evolution module summarizes the effect of the execution results of this task through self-evaluation and ground uplink evaluation, summarizes and stores the "numerical" experience, which can be directly used for the next similar task. The intelligent semantic interpretation system performs semantic interpretation of the information in high-resolution images, completes the image data process, and directly distributes the simplified data information to the user's handheld devices and command center in a "point-to-point" manner as needed and in a timely manner through onboard communication means, which comprehensively improves the timeliness from the process of information processing, distribution, and utilization, provides users with fast near-real-time information, improves the ability to perceive emergencies, and provides effective space-based information support for the command center [15].

9.6 Intelligent Semantic Interpretation System

Intelligent semantic interpretation system is to complete the process of image digitization through information evaluation, image fusion, and data compression in orbit and directly distributes the simplified data information to the user's handheld device and command center in a "point-to-point" manner through onboard communication means to provide users with visual real-time data information and improve their ability to perceive emergencies.

In image semantic interpretation, data characteristics and characteristics are analyzed first, then features are extracted, and the relationship between features and semantics is built through a one-to-one mapping relationship to generate semantic content and features. Finally, learning and training are conducted according to the relationship between semantic content and features, and a knowledge-based semantic model is established, and finally the image interpretation results using semantic representation are obtained. This method can effectively improve the association relationship between different features, provide more association concepts for the mining of potential knowledge, make the understanding of target objects more comprehensive and accurate, and achieve effective information mining and intelligent interpretation of massive data [16].

With the continuous improvement of the resolution of the payload satellite, the amount of data generated by the payload satellite shows a geometric progression growth trend. The ground resolution of the visible spectrum segment of the optical payload has reached several meters or even less than 1m, the number of hyperspectral spectrum segments has reached hundreds, the dynamic range and radiation resolution have continuously improved, and the number of pixel quantization bits has increased, making the original data rate of the spaceborne payload reach several or even dozens of Gbps. This creates a huge contradiction with the limited transmission and storage resources on the payload platform, making the effective processing of massive data on the satellite becomes an urgent problem to be solved in the development of payload technology.

9.7 Intelligent Load Onboard Intelligent Processing Technology Scheme

Users in different application fields have different requirements for load information, but the data products provided by the current processing mode to users are basically simple load image data, not the products required by users. The development of onboard intelligent processing can produce products in real-time according to user needs on the orbit, and distribute the products directly to different users, so that users can obtain the required data through simple operations, which will greatly expand the application market in the load field. At present, there is a contradiction between "more and less" in satellite data. On the one hand, the ground system obtains a large amount of earth observation data every day. On the other hand, due to the backward information processing technology, a large number of data cannot be processed in time, resulting in great waste. The development of onboard intelligent processing can effectively and automatically process and extract information from a large number of earth observation data. At the same time, the ground station and receiving equipment will be miniaturized and simplified, so as to realize the transformation from the data service of the existing system to the fast information service. Intelligent load satellite needs to provide personalized and diversified information services for the public through intelligent processing of onboard software or combined with ground processing. The recognition technology of load image has been widely used on the ground, such as real-time data statistics technology of sports events, intelligent monitoring technology, and automatic driving technology. These technologies have certain similarities and can be used as a reference for intelligent load satellite image processing. In addition, the gradual maturity of machine deep learning technology, the rapid development of high-performance computing platforms such as multi-core CPU (central processing unit) + multi-core GPU (image processing unit), and the support of big data and cloud computing technology all provide technical support for onboard processing of intelligent payload satellites [17].

The method based on depth learning can be widely used in other photogrammetry and remote sensing data processing, in addition to being effectively used in remote sensing image classification and target retrieval. For example, Hu Xiangyun and others used the deep learning method to process the LIDAR point cloud data. Point cloud data filtering in mountainous forest areas, extracting digital elevation models from point cloud data is difficult to achieve automation, generally requires human–computer interaction, and consumes a lot of manpower and material resources. At present, the team uses machine learning methods to train and learn knowledge in the process of human–computer interaction and then applies it to the automatic processing of point cloud data. The accuracy rate of automatic processing reaches more than 95%, greatly improving the operation efficiency [18].

Compared with traditional imaging methods, the imaging method of satellite payload system based on artificial intelligence technology has the following advantages:

1) On-demand imaging, more targeted.
2) Low-resolution perception ability, high-resolution key target detailed inspection ability.
3) Reduce the amount of downlink data and the pressure of satellite information transmission.
4) High timeliness of information utilization.
5) It has the function of learning and evolution.

Satellite payload system based on artificial intelligence technology can systematically design the whole link of intelligence collection, processing, distribution, and application from the source of data acquisition, improve the timeliness and usefulness of information, and provide effective support for users to obtain information on demand and in time. It can be applied to large-scale load imaging, automatic search, identification and positioning of interested targets, tracking and detailed investigation of dynamic and static targets, and has independent situation awareness and intelligence processing capabilities. It has great application value in the field of real-time response in orbit, rapid positioning and interpretation of emergencies, and rapid capture and tracking of moving targets.

9.8 Digital Multi-function Load

The development trend of satellite system is multi-function, networking, and intelligence. It integrates communication, intelligence reconnaissance, target detection and early warning, information distribution and information countermeasures, and forms a network with each other to achieve real-time acquisition and fusion of multi-dimensional information. The load based on software definition has the characteristics of generality, standardization, and modularization, which can realize the unified architecture of multi-function system and realize different functions through software definition and reconstruction [19].

9.8.1 American SCAN Test Bench

The NASA Space Communication and Navigation Test Platform (SCAN) is the first experimental communication and navigation system installed on the International Space Station. It was launched into the International Space Station (ISS) by the Japanese transfer vehicle in 2012. Its development purpose is to evaluate the application efficiency of SDR in NASA space missions. The test items include: software application of communication, navigation, and network based on SDR architecture, including testing waveform with communication rate up to 800Mbps, GPS positioning software, on-orbit routing technology based on space standards and commercial IP protocol, delay tolerance network (DTN) security, frequency and time transmission, and realization of link data frame generation, remote control, and telemetry, synchronization and series of standards such as encoding and pseudo-random data generation, as well as orbit acquisition, link automatic operation and adaptive cognitive applications [20].

The SCAN test bed operates in NASA's space and ground networks, including the relay satellite (TDRS) and multiple ground stations.

The S-band and Ka-band radio systems on the test platform can communicate with the Baisha ground station through the relay satellite, and the S-band radio system can also directly communicate with the ground station. Among them, the SCAN test bed receives the remote-control command via the RF system on the ISS and sends telemetry data to the ground network, which then sends the data to the control center at the Green Research Center for real-time processing. During the test, the ground computer can view the telemetry and observe the SDR load performance. The radio frequency system of SCAN test bench can communicate with any S-band ground station authorized to receive its signal.

The space demonstration and verification tasks of SCAN test bench include:

1) Validate networked communication and navigation technology. Multi-hop delay communication service based on network protocol and store-and-forward technology.

   It is helpful to alleviate the problems of interruption, high delay, and insufficient coverage for all users in the current inter-satellite communication. At present, network technology has been widely used in existing ground systems, such as the high-speed transmission of spatial link information flow and frame data using the ground infrastructure. However, the existing space-based systems have not realized large-scale networking. This problem can be solved by deploying the SDR platform to enhance the waveform generation capability and provide technical means for the future interconnection and cooperation of space networks.

2) Validate high-speed data transmission rate for robot and manned exploration missions. The massive data collected by sensors and equipment in scientific and exploration missions put forward new requirements for high-speed data transmission of satellites, such as high-definition video and synthetic aperture radar data. Even after compression, the amount of scientific experiment data is considerable. SDR provides flexibility and scalability for realizing and controlling multi-channel and multi-function data transmission.

3) Validate international interoperability data communication protocols for future space exploration missions. At present, the communication interfaces of space network (SN), near-earth network (NEN), and deep space network (DSN) are relatively complex, which requires that the space vehicle must have a very flexible and reconfigurable radio and data processing system to support the existing multiple network operating modes and future international standard operating modes. SDR could generate multiple waveforms and can be reconfigured according to existing and future requirements.

SCAN test bench mainly comprises GD, JPL, and Harris SDRs. Among them, GD SDR was developed by GD Company, a NASA partner, and is an S-band communication system compatible with Tracking and Data Relay Satellite (TDRSS) and STRS, which is used to develop the fourth generation of TDRSS transponder. JPL SDR has a full duplex, S-band communication function compatible with TDRSS and STRS, and GPS navigation receiver function. S-band and GPS can work at the same time, aiming to verify the multi-zone radio navigation system.

The above two S-band radios can communicate through the medium gain antenna (MGA) of the universal joint, the low gain antenna (LGA) pointing to the space, or the LGA pointing to the ground. Harris SDR is NASA's first in-orbit Ka-band transceiver (compatible with TDRSS) for space operation applications. It can communicate with the relay satellite through the space-facing universal joint high-gain antenna (HGA). Harris has completed the system design, manufacturing, testing, and delivery in only 14 months.

9.8.2 UK Full Digital Payload

OneWeb and SatixFy UK announced plans to add a digital technology verification payload to OneWeb's 2021 launch plan, that is, a full digital payload. This new technology will lay the foundation for developing more flexible satellites to effectively support the peak demand without making the constellation too large.

This full digital payload is a digital transparent payload, which will verify the full beam hopping capability in the forward link and the return link. It includes a satellite processing subsystem, which enables users and ground gateway links to operate independently and have different capacities. Electronically controlled multi-beam antenna has the true time delay beamforming ability and can point and switch multiple beams to multiple directions at the same time, which is a major feature of the increase.

The combination of these characteristics significantly improves the flux of the designated area within the satellite coverage, while continuing to meet all national security needs. The capacity dynamically allocated to active regions is much higher than that of inactive regions (regions that do not need so much bandwidth). The inactive regions are only scanned to measure demand. The beam hopping capability also supports the seamless switching of mobile equipment between beams and satellites. One example is the aviation terminal used for in-flight communication, which can communicate with LEO and GEO satellites at the same time and maintain a make-before-break connection between the upper and lower satellites. The maximum capacity of multiple sources can be directed to busy airports and other hot spots.

The European Space Agency and the British Space Agency are conducting in-depth discussions on cooperation and support.

Joel Gart, CEO of SatixFy Group, said: "We have designed several sets of chips in the whole satellite value chain, including terminals, payloads and gateways, and created a complete ecosystem based on our own software. The new OneWeb satellite will demonstrate all these capabilities."

Massimiliano Radovaz, the chief technology officer of OneWeb, said: "This cutting-edge satellite will be put into full use, which is an excellent opportunity to demonstrate our payload technology. The improvement of performance, efficiency, and target capacity, coupled with advanced ground infrastructure, make us at the forefront of low-orbit communication network services."

OneWeb launched the first satellite in 2019, and soon began to implement the regular launch plan, providing global commercial broadband services by the end of 2021.

9.8.3   The European Space Agency's Second-Generation Galileo Satellite

According to the report on the GPS World website on August 14, 2020, after the European Commission proposed to accelerate the development of the next generation of Galileo satellites, the European Space Agency (ESA) conducted a bid for the first batch of the second-generation Galileo satellites to European satellite manufacturers. In addition to all the services and capabilities provided by the current first-generation satellites, the second-generation Galileo satellites will also make major improvements to provide new services and capabilities. The first satellites are expected to be launched in 2024, and the ground station will also be upgraded synchronously.

ESA has had a competitive dialogue with three large system integrators, Airbus, OH System AG, and Thales Alenia Aerospace, for nearly 24 months. On August 11, ESA issued an invitation for "best and final quotation" to the three companies. ESA is adopting a dual-source procurement strategy. At the end of 2020, two companies were selected from the current three bidders to sign two parallel contracts. According to the development requirements, the two companies manufactured two satellites, up to 12 satellites.

Main features of the second-generation Galileo satellite:

On-orbit reconfiguration: The second-generation Galileo satellite has digital design and full flexibility and can be reconfigured in orbit to meet the expected development needs of end users.

Advanced navigation antenna: ESA has completed a lot of research and development work on advanced navigation antenna and looks forward to further technical development. In order to ensure feasibility, ESA has built a similar antenna at its European Space Technology Research Center (ESTEC) in the Netherlands as a proof of concept and shared advanced technology with three bidders.

Flexible payloads: Fully flexible payloads are also a serious challenge, and there are no such navigation satellites in operation at present.

The second-generation Galileo satellite was previously known as the "transition batch" satellite, which was named after the temporary upgrade of the satellite to solve the potential risks caused by the delay in delivery of the second-generation Galileo satellite. This renaming reflects the hope of the European Commission and EU member states to further improve the technical capabilities of the Galileo system. The second-generation Galileo satellite gradually replaces the current first-generation satellite, forming a complete constellation with the standby satellite in orbit.

References

1 University of Cincinnati (2016). New artificial intelligence beats tactical experts in combat simulation. https://www.sciencedaily.com/releases/2016/06/160627125140.html (accessed 27 June 2016).

2 Norton, C.D., Pellegrino, S., Johnson, M. (2013). Findings of the KECK Institute for Space Studies Program on small satellites: a revolution in space science. *Proceedings of the 27th Annual AIAA/USU Conference on Small Satellites*. Logan, UT (August 2013): NASA.

3 Liewer, C., Klesh, T., Lo, W. et al. (2014). A fractionated space weather base at L5 using CubeSats and solar sails. *Advances in Solar Sailing.* 2014: 269–288.

4 Kiran, M. and Michael, S. (2014). SkySat-1: very high-resolution Imagery from a small satellite. In: *Sensors, Systems, and Next-generation Satellites XVIII*. Bellingham: SPIE.

5 Zhaocai, Z. and Luqing, Z. (2015). Latest developments in research on earth observation small satellites. *Space International* 2015 (11): 46–51.

6 Bing, Z. (2011). Intelligent remote sensing satellite system. *Journal of Remote Sensing* 15 (3): 415–43l.

7 Zhaocai, Z. (2015). Commercial company of earth observation accelerating "data democracy". *Satellite Application* 2015 (3): 67–68.

8 Zhou, Y., Wang Peng, F., and Danying. (2015). System innovation and enlightenment of SkySat. *Spacecraft Engineering* 24 (5): 91–98.

9 Stanley, D. and Christina, D. (2010). Tactical Satellite-3 mission overview and initial lessons learned. *24th Annual AIAA/USU Conference on Small Satellites*. Washington D.C. (August 2010): AIAA.

10 CCSDS (2014). *CCSDS 130.0-G-3: Overview of Space Communication Protocols*. Washington D.C.: CCSDS.

11 CCSDS (2012). *CCSDS 702.1-B-1: IP Over CCSDS Space Links*. Washington D.C.: CCSDS.

12 Deren, L. and Xin, S. (2005). The research of intelligentize a observation of ground. *The Science of Topography* 30 (4): 9–11.

13 Junwei, W. (2007). *The Research of Harbor Detection in Remote Sensing Image*. Xi'an: Xi'an Electronic Science and Technology University.

14 Su, Y., Sun, Q., Jiao, J. et al. (2012). Study on the advanced remote sensing technology "intelligent eyes". *The Proceedings of 1st Conference on High Resolution Earth Observation*, Beijing (December 2012).

15 Elnagar, A. and Gupta, K. (1998). Motion prediction of moving objects based on autoregressive model. *IEEE Transactions on Systems, Man and Cybernetics* 28 (6): 803–814.

16 Oliva, A. and Torralba, A. (2006). Building the gist of a scene: the role of global image features in recognition. *Progress in Brain Research: Visual Perception* 155: 23–26.

17 The SCaN Testbed (2012). The US Scan testbed will develop a new generation of space communications technology at the International Space Station. *Spacecraft Engineering* 21 (2): 78.

18 Richard, C., Reinhart, J.S., Sandra, K.J. (2014). Recent successes and future plans for NASA's space communications and navigation testbed on the international space station. *65th International Astronautical Congress*, Toronto, Canada (2 October 2014).

19 Johnson, S.K., Reinhart, R.C., and Kacpura, T.J. (2012). CoNNeCT's approach for the development of three software defined radios for space application. *Proc. IEEE Aerospace Conference*, Big Sky, Montana (3–10 March 2012), pp. 1–13.

20 Chelmins, D., Mortensen, D.J., and Shalkhauser, M.J. (2014). Lessons learned in the first year operating software defined radios in space. *AIAA SPACE 2014 Conference and Exposition*, Pasadena, CA (4–7 August 2014), pp. 1–11.

10

Future Development of Intelligent Satellite

Artificial intelligence is a cutting-edge technology in the field of information. A series of recent technological breakthroughs have given this field a strong impetus and are moving rapidly toward a new stage of development. The United States, Japan, Europe, and other developed countries and regions have reached a high consensus on this issue and have recently launched a series of new measures in a row to seize the lead in the upcoming new round of scientific and technological revolution.

At present, the international space power led by NASA has taken space as an important stage for AI to play its role. Many space tasks that have been carried out or will be carried out have adopted AI technology to improve the efficiency of related tasks. Although the current application of AI technology in space is still limited, and the achievements are not outstanding enough, the power of AI has been demonstrated, and the future development direction it represents has also begun to emerge.

10.1   Application Prospect of AI in Aerospace Field

10.1.1   Autonomous Operation Satellite System Based on Artificial Intelligence

The satellite is generally composed of multiple subsystems, which have relatively independent functions. At the same time, they must cooperate with each other to effectively complete the task.

In April 1999, NASA conducted an autonomous control test on Deep Space1 (DS1). Its control structure is shown in Figure 10.1. The main body of the autonomous control is the Remote Agent, which coordinates with the ground control system, satellite real-time control software and hardware, and sensors to complete the control of the satellite. The remote agent consists of three parts, namely, planning and scheduling, intelligent execution agent, pattern recognition and fault

Intelligent Satellite Design and Implementation, First Edition. Jianjun Zhang and Jing Li.
© 2024 The Institute of Electrical and Electronics Engineers, Inc. Published 2024 by John Wiley & Sons, Inc.

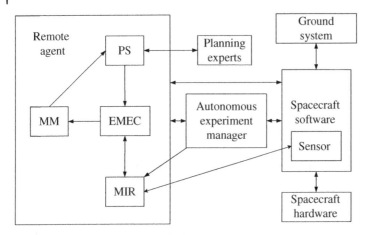

Figure 10.1 Autonomous control structure of Shenkong 1.

recovery, and a task manager. At work, the planning and scheduling module decomposes the classified tasks into high-level plans and then submits the plans to the intelligent execution agent. The intelligent execution agent module further decomposes the plans into detailed control instructions and then sends them to the satellite software to complete the tasks. NASA gave another autonomous operating system architecture in the ST7 plan. It is composed of three major parts, the top of which is system-level autonomy. It is mainly responsible for system-level planning and scheduling, as well as system-level state detection, fault diagnosis, and system-level reconstruction. This layer is responsible for overall decision-making and coordination. The second part is the independent satellite subsystem of brother. These subsystems have certain independent functions and can complete their own tasks and monitor their own status. The third part is the autonomous science subsystem, which is responsible for managing each payload, independently analyzing the data obtained together and guiding the next work [1].

The quality of the autonomous operation architecture is the key to the effective work of the autonomous operation system. In general, the satellite autonomous operation architecture tends to adopt the distributed multi-agent control structure with central control, in which the master agent is responsible for system-level planning and control, and each other subsystem is controlled by a corresponding subagent [2].

10.1.2 Satellite State Detection, Fault Diagnosis, and System Reconstruction

With the rapid development of artificial intelligence technology, a variety of artificial intelligence fault diagnosis methods such as expert system, fault tree, neural network, and so on have emerged. For the expert system fault diagnosis method, the system is comprehensively reasoned by using various rules, and the output information of the operating system is compared with the experience information provided by the user, so as to find the existing or possible faults, and finally verify the correctness of the diagnosis results through the user. According to the fuzzy theory, a fuzzy fault diagnosis method is designed can locate the fault and predict the impending fault in advance.

In recent years, the research of fault tree model to find fault sources in complex systems has attracted people's attention. The system fault and fault cause is drawn into a fault tree, which can reflect the direct relationship between the fault and each cause. By studying the fault diagnosis method based on neural network, the neural network is used as a classifier and combined with the advantages of other diagnosis methods to carry out fault diagnosis. The fault diagnosis methods based on information fusion, such as BP neural network, Bayes theory, fuzzy algorithm, D-S reasoning theory, and neural network, adopt information fusion theory for fault diagnosis and improve the accuracy of fault diagnosis [3].

The fault diagnosis method of artificial intelligence takes human factors into full consideration, and the inference of actual systems, especially nonlinear systems and complex systems, is reasonable. It is a kind of potential fault diagnosis method.

10.1.3 Satellite Intelligent Autonomous Control Technology

The satellite intelligent control includes two modes: the large loop of space and earth and the on-orbit mode. The intelligent autonomous control is to use intelligent technology to realize the function of on-orbit autonomous control. With the development of computer and artificial intelligence, Fu Jing, an American Chinese, first proposed to develop the concept of intelligent control by combining artificial intelligence with automatic control. If artificial intelligence is a powerful means to realize the highly autonomous operation of the system, then intelligent control is the only way to realize the autonomous operation of the control system. Academician Yang Jiaxi put forward several key areas of space intelligent control in 1995: control of flexible structure space vehicles, autonomous navigation and orbit control, fault targeting and system reconstruction, etc. Academician Wu Hongxin discussed the key scientific issues involved in satellite intelligent control from the perspective of the development of intelligent control science in 2001.

To seize the highest point in space, the major aerospace countries and powers have increased their investment in intelligent autonomous control technology in their own space plans. For example, ESA has been aiming to catch up with and surpass the United States and the former Soviet Union with the help of advanced unmanned autonomous control technology in orbit since the 1970s. NASA identified eight key technologies including artificial intelligence, intelligent control and robust multivariable adaptive control in the "Space Policy and Plan" formulated in 1988. In 1992, NASA invited bids worldwide for research on intelligent control technology and its application in satellite orbit and attitude control. The U.S. Air Force's 2025 plan, the U.S. Aerospace Command's 2020 long-term plan and NASA's New Prosperity Plan all put intelligent autonomous technology at the top of the list. One of the goals of the New Prosperity Plan is to develop autonomous satellites, and the autonomous technology is listed as a key point in the development of the plan. The purpose is to enable the satellite to independently complete the guidance, navigation, and control (GNC), data processing, fault diagnosis, and partial reconstruction and maintenance work, thus greatly reducing the dependence on ground measurement and control, communications, and other support systems [4].

10.1.4 Intelligent Planning and Scheduling of Satellites

In the future, the satellite system will become larger, more complex, more control activities, tight measurement and control resources, and high reliability and security requirements. In the face of such a mission, how to design the flight process of the satellite, how to determine the control operation of the satellite, how to formulate the flight control plan and adjust it in real-time and dynamically in the mission, and what mode to adopt to ensure the success of the flight control mission are all the basic problems that need to be solved in the space flight mission planning.

The object of mission planning is the flight control process, and the goal is to achieve a feasible or optimal planning process based on experience and knowledge under certain resource constraints. Space mission planning is a typical knowledge-processing procedure involving complex logical reasoning and many constraints. This problem is suitable to be solved by the theory of artificial intelligence.

In 2000, NASA applied the autonomous planning program to the scheduling of control satellites for the first time. The program generated the plan according to the high-level objectives formulated on the ground and monitored the implementation, detection, diagnosis, and recovery in case of problems. The follow-up MAPGEN carried out daily planning for NASA's Mars exploration, and MEXAR2 carried out mission planning for ESA's Mars Express mission in 2008 [5].

10.1.5 Satellite Collaboration and Cooperation

In future satellite missions, communication, cooperation, and cooperation between satellites are required to complete specific complex tasks. If such cooperation and communication require receiving signals, processing, planning, and sending instructions through the ground, it will not only reduce efficiency but also easily miss the best operation time. In addition, the autonomous system and cooperation between satellites can also reduce the use of onboard resources, such as power resources and communication resources. Of course, if we want to achieve autonomous cooperation and cooperation between satellites, we need very large onboard storage equipment to track the satellite interaction in real time, which is also a difficult point to be solved.

The most common form of satellite collaboration is constellation, and the more complex form of collaboration is satellite cluster system. The satellites in the cluster have different spatial distribution, payloads, functions, etc. By complementing each other's advantages, they cooperate to complete the established tasks under complex constraints. Timeliness, cluster resource planning, system on-orbit health and safety, system unsupervised operation, system fault diagnosis, and reconstruction will be a huge and complex project if they rely on the ground system for processing. At this time, it will be an inevitable way to establish an intelligent satellite cluster with autonomous operation capability and realize the highly autonomous operation capability of cluster satellites.

The need for autonomous satellite cooperation in satellite missions is very urgent. Intelligent cooperation focuses on how to cooperate with people and intelligent carriers (computer agents) to achieve common goals. NASA briefly gives several aspects that require artificial intelligence [6].

10.1.5.1 Lots of Scientific Data

In the past few decades, the rate of scientific data collected and transmitted by satellites has increased by several orders of magnitude. This is due to the development of sensor and computer technology. At the same time, the need for data processing systems to process these data has become very urgent, but the traditional data processing methods have not changed much in recent decades.

10.1.5.2 Complex Scientific Instruments

With the increasing complexity of space missions and loads, the number of scientific instruments on the satellite has soared, so it is very difficult to make full use of the effective resources on the satellite. The complexity of equipment will lead to a surge in the demand for onboard resources and planning constraints, so it is difficult to make full use of all instruments under limited resources. Experience has shown that it is of great significance to fully coordinate instruments and space resources with AI to complete complex space tasks.

10.1.5.3 Increased Number of Satellites

In the past decade, the number of satellites has increased rapidly. The recent widespread attention to small satellites will accelerate this growth rate. To maximize the potential of orbiting satellites, NASA uses different models for applications: constellation, formation, or constellation. However, to prevent the complex ground system from emerging in response to many satellites, a virtual platform must be introduced to manage multiple satellites. Many satellites are built into a virtual platform of the system by the intelligent system micro-core, which can make the operation of the system more flexible and efficient.

10.1.6 Multi-satellite Mission Management

Cluster satellites can accomplish tasks that could not be accomplished by a single satellite in the past and have many advantages [7].

1) It can prevent single point (single system or equipment) failure in the system from causing the failure of the whole task.
2) It can observe and monitor a single target or multiple targets in different forms and different positions at the same time (or use many small antennas to form a large antenna system).
   Reduce the complexity of satellites by simplifying the number of equipment and supporting systems.
3) Replace or add equipment (functions) by adding new satellites in constellation or cluster

NASA Microwave Anisotropy Detector was launched in 2001, requiring four people to operate the entire system. Because there is only one satellite, it requires fewer people to operate. Iridium satellite has a total of 66 satellites. At first, about 200 people were needed to complete the basic operations; that is, about three people were needed for each satellite. NASA's GlobalStar satellite system comprises 48 satellites in total, requiring about 100 people to carry out daily operations. In fact, both Iridium and Global Satellite Systems are composed of satellites with similar functions. Compared with systems composed of different satellites, their operation, and maintenance are also relatively simple, but the development of satellites in the future will be more complex.

To save the operation cost on the ground, it is an effective way to use artificial intelligence to realize satellite autonomy. For example, it is usually necessary to transmit all the data on the satellite to the ground for backup and analysis. The amount of data generated by the early equipment is small. Due to the abundant resources on the satellite, the data transmission is not a big problem. However, with the development of technology, the amount of data generated by the relevant equipment has increased sharply, and due to the mission requirements, many satellites operate in high orbit or interstellar orbit, which requires higher power and gain antennas to complete the transmission of these data, which not only needs to increase the workload of ground operators but also will significantly increase the launch cost. Therefore, the analysis, identification, and interpretation of relevant data on the satellite can be completed through the intelligent system, which can effectively improve the efficiency of onboard equipment, reduce information leakage and increase the anti-interference ability of the satellite [8].

10.2 The Next Development Gocus

10.2.1 Design of Onboard Intelligent Chip

With the increasing development of AI algorithms and application technologies and the gradual maturity of the industrial environment of AI special chip ASIC, AI ASIC will become the inevitable trend of AI computing chip development. The AI application market is huge, and the accumulation of remote sensing data, telemetry and telecontrol, communication and navigation has formed a massive scale, which provides a huge space for the development of artificial intelligence chips on the satellite [9].

10.2.2 Satellite System Design Based on Artificial Intelligence

In the past ten years, the rapid development of ground commercial electronic system technology has made the electronic components selected in the early stage obsolete when they were in service. The electronic system should no longer be built based on specific electronic components during design but provide a low-cost, scalable system that supports the improvement and upgrading of the whole life cycle of the project through reasonable architecture design. Thus it can simplify the cost of design, development, testing, integration, maintenance and upgrading. Therefore, the following design objectives can be proposed for the research and development of electronic systems [10].

1) Based on the open system architecture, the processing, communication and computing resources that constitute the core of the system can be flexibly expanded and reconfigured according to the system requirements.
2) Based on existing commercial standards or mature products, third parties can participate in the software and hardware research and development of space-borne electronic systems to reduce the development cost.
3) Through spatiotemporal isolation technology, it supports local modification and upgrading of the system and reduces the cost of system update and re-authentication. At the same time, the fault is blocked locally to improve the overall reliability of the system.
4) Carry out research on the design method of electronic system based on artificial intelligence, use artificial intelligence knowledge to complete the processing of various instructions and telemetry, and realize a processing platform that can meet the flexible expansion of multiple tasks and support the flexible reconstruction of system resources in case of failure.

10.2.3 On-track Fault Detection and Maintenance Based on Artificial Intelligence

The satellite system is a dynamic system. It needs to monitor the working state when it is in orbit. It also needs to diagnose when there is a fault. In addition, each important system of the satellite has taken redundant backup measures, hoping to recover its function through reconstruction in case of failure. State detection, fault diagnosis, and system reconstruction are important components of satellite autonomous operation. The realization of intelligence and autonomy can improve the stability and viability of satellite on-orbit operation. The pattern recognition and system reconstruction system adopts the pattern-based fault diagnosis method, composed of two parts: pattern recognition (MI) and system reconstruction (MR). Pattern recognition is the sensing part of autonomous control, which is responsible for tracking state changes. During operation, the pattern recognition module monitors the control commands sent by the intelligent actuator to the satellite, uses its own satellite model to infer the state of the satellite when executing these commands, and then compares it with the information collected by the sensor. If the collected information is consistent with the normal command, it considers that the command has been executed correctly. If not, the satellite is in an abnormal state. When a fault occurs, it is necessary to identify specific firmware components and failure modes based on sensor information and reasoning results [2].

In addition to pattern recognition, the system should also solve faults – reconfiguration. For example, the main engine of satellite has many redundant valves. In case of operation or failure, different valves can be selected for configuration, which has different costs. The task of the system reconfiguration module is to complete the configuration according to the minimum cost when the configuration requirements are proposed. When a fault occurs, it can be recovered from the fault by repairing or looking for reconstruction. When the fault cannot be eliminated, enter the standby mode.

The realization of intelligent state detection and system reconstruction is the development direction of future satellites. Its technology should be universal and can be quickly extended to different platforms [11].

Based on the comprehensive analysis of the research status of satellite fault diagnosis and fault tolerance technology, the multi-satellite, multi-station, and multi-level comprehensive diagnosis technology, the intelligent fault diagnosis technology of multi-method fusion, and the timely and effective fault tolerance technology will become the development direction of satellite fault research in the future [12].

1) **Multi-satellite, multi-station, and multi-level comprehensive diagnosis technology**: In view of a large number of satellite ground station equipment, a large amount of satellite observation data, and the different data quality of each station, the potential failure modes of each component of the

system are studied hierarchically using the fault tree, and the fault modeling and theoretical analysis of hard fault, soft fault, or multi-fault and composite fault are carried out based on the fault generation mechanism and response characteristics. Then a multi-satellite, multi-station, and multi-level comprehensive diagnosis method is proposed, which can not only avoid misjudgment of system status caused by single station data fault but also detect soft fault timely and effectively, improve fault processing efficiency, and realize rapid discovery of abnormal problems.

2) **Multi-method integrated intelligent fault diagnosis technology**: In satellite systems, the use of a single diagnosis method often fails to meet the needs of system fault diagnosis. Considering various factors such as system function, effect, and objective conditions, the analytical model, signal processing, artificial intelligence, and other fault diagnosis methods are combined to define the intelligent diagnosis strategy. Combined with a multi-level diagnosis strategy, appropriate diagnosis granularity, model signal, and intelligent knowledge are combined to learn from each other so as to realize rapid and accurate fault diagnosis and make the whole system reach a higher level of intelligence.

3) **Timely and effective fault tolerance technology**: Timely and effective fault tolerance technology in the satellite system enables the navigation system to have self-monitoring capability. Under fault diagnosis conditions, the fault equipment can be isolated in time and the rest equipment can be reconstructed, so that the satellite system can operate normally or safely with reduced performance. Therefore, the intelligent fault diagnosis technology combined with multi-data and multi-method fusion can automatically mobilize information resources according to its own fault system, adopt active reconfiguration for hardware faults, and adopt passive and robust hybrid intelligent and automatic fault-tolerant technology for software faults, so as to eliminate the impact of faults on the system in a timely and effective manner and ensure the healthy operation of the system [13].

10.2.4 Satellite Intelligent Control Based on Artificial Intelligence

The satellite intelligent control includes: the large loop of space and earth and the on-orbit mode. With the development of computer and artificial intelligence, satellite intelligence technology is developed by combining artificial intelligence with automatic control of satellite intelligence.

The next step is to increase the investment in intelligent autonomous control technology, combine it with artificial intelligence technology, and develop advanced unmanned autonomous control technology in orbit, so that satellites can independently complete guidance, navigation and control (GNC), data processing, fault diagnosis, and partial reconstruction and maintenance, thus greatly reducing the dependence on ground measurement and control, communication and other support systems. The application of real-time intelligent autonomous attitude control, intelligent autonomous GNC, and intelligent information technology in aerospace control systems, platforms, and payloads will be realized [14].

10.2.5 Space-ground Integration Based on Artificial Intelligence

With the promotion of the major project of the space-ground integrated information network, the ideas about the construction of space-based information systems are gradually becoming clear. However, traditional space-based information systems have poor in-orbit information processing capabilities and cannot meet the diverse space-based information service requirements such as high time sensitivity and multi-task coordination.

According to the requirements of space-based information for high real-time, diversified, and systematic applications, combined with artificial intelligence technology, through inter-satellite link connection, a virtual large satellite is formed to provide users with high-performance and efficient spatial information processing services, which has high system flexibility and survivability and realizes comprehensive perception, information aggregation processing, high-speed distribution, network management, security protection, and other functions. It is the spatial form of the ground-integrated data center, and it will effectively break through the constraints of the large-scale characteristics of space and space-time on the timeliness of information and innovate the space-based information service model

10.2.6 Satellite Intelligent Platform Based on Artificial Intelligence

With the development and improvement of intelligent learning algorithms, learning algorithms have developed vigorously in various fields of industry. From the perspective of satellite platform, the information rate of telemetry and remote-control data is from 2 kbps to 1 Mbps. The amount of information transmitted by a single satellite data is close to Tb per day. In fact, the amount of telemetry information of each subsystem within the satellite is as high as Pb. From the perspective of big data, how to decompose and predict effective intelligence from huge amounts of telemetry and telecontrol information, or achieve efficient control of satellite management through existing historical information, or even autonomous control in unknown emergencies, is an effective way to improve the usability and ease of use of satellite platforms.

Therefore, the development direction of intelligent satellite platform is to combine intelligent learning with satellite system design and satellite intelligent platform, and finally achieve the goals of intelligent design of satellite system, intelligent fault detection and maintenance of satellite in orbit, and intelligent control of satellite on intelligent chip.

10.3 Summary

In a series of processes such as satellite development, testing, flight control, delivery, and use, the problems of the unattended space environment, the high cost of testing and maintenance, and many factors of fault problems have been puzzling scientific researchers. Artificial intelligence supporting satellite system technology is a powerful means to solve these problems and is one of the development directions of satellite platform design in the next decade. In the future, it will not only be able to process complete information but also process incomplete information, and even intelligently supplement incomplete information, and make the processing of information and data more mature, efficient, and accurate according to the feedback system. At the same time, experience is constantly accumulated in daily operation, so that the AI system can adapt to the changing environment, gradually realize the automatic evolution mechanism, make the AI system itself constantly learn, change the single passive processing information into active, intelligent processing information, and even have a certain predictive ability.

References

1 Mnih, V., Kavukcuoglu, K., Silver, D. et al. (2015). Human-level control through deep reinforcement learning. *Nature* 518 (7540): 529–533.

2 Zhang, M., McCarthy, Z., Finn, C. et al. (2016). Learning deep neural network policies with continuous memory states. *2016 IEEE International Conference on Robotics and Automation (ICRA)* (16–21 May 2016), IEEE, pp. 520–527.

3 Schmidhuber, J. (2015). Deep learning in neural networks: an overview. *Neural Networks* 61: 85–117.

4 Sutton, R.S. and Barto, A.G. (1998). *Reinforcement Learning: An Introduction.* Cambridge: MIT Press.

5 Arulkumaran, K., Deisenroth, M.P., Brundage, M. et al. (2017). A brief survey of deep reinforcement learning. *IEEE Signal Processing Magazine* 34 (6): 26–38.

6 Barto, A.G., Sutton, R.S., and Anderson, C.W. (1983). Neuronlike adaptive elements that can solve difficult learning control problems. *IEEE Transactions on Systems, Man, and Cybernetics* 1983 (5): 834–846.

7 Zhu, Y., Mottaghi, R., Kolve, E. et al. (2017). Target-driven visual navigation in indoor scenes using deep reinforcement learning. *2017 IEEE International Conference on Robotics and Automation (ICRA)* (29 May–03 June 2017), Singapore: IEEE, pp. 3357–3364.

8 Zhang, J., Springenberg, J.T., Boedecker, J., and Burgard, W. (2017). Deep reinforcement learning with successor features for navigation across similar environments. *2017 IEEE/RSJ International Conference on Intelligent Robots and Systems (IROS)*, Vancouver, BC, Canada (24–28 September 2017), pp. 2371–2378.

9 Tai, L., Paolo, G., and Liu. M. (2017). Virtual-to-real deep reinforcement learning: continuous control of mobile robots for mapless navigation. *2017 IEEE/RSJ International Conference on Intelligent Robots and Systems (IROS)*, Vancouver, BC, Canada (24–28 September 2017), pp. 31–36.

10 Parisotto, E., Chaplot, D.S., Zhang, J., and Salakhutdinov, R. (2018). Global pose estimation with an attention-based recurrent network. *2018 IEEE/CVF Conference on Computer Vision and Pattern Recognition Workshops (CVPRW)*, Salt Lake City, UT, USA (18–22 June 2018), pp. 350–359.

11 Gu, S., Holly, E., and Lillicrap, T. et al. (2017). Deep reinforcement learning for robotic manipulation with asynchronous off-policy updates. *2017 IEEE International Conference on Robotics and Automation (ICRA)*, Singapore, pp. 3389–3396.

12 Levine, S., Pastor, P., Krizhevsky, A. et al. (2016). Learning hand-eye coordination for robotic grasping with deep learning and large-scale data collection. *The International Journal of Robotics Research* 0278364917710318.

13 Finn, C., Tan, X.Y., Duan, Y. et al. (2016). Deep spatial autoencoders for visuomotor learning. *2016 IEEE International Conference on Robotics and Automation (ICRA)*, IEEE, pp. 512–519.

14 Riedmiller, M., Hafner, R., Lampe, T. et al. (2018). Learning by playing - solving sparse reward tasks from scratch. *Proceedings of the 35th International Conference on Machine Learning, PMLR Stockholm*, Sweden (10–15 July 2018), 80, pp. 4344–4353.

Index

Intelligent Satellite Design and Implementation, First Edition. Jianjun Zhang and Jing Li.
© 2024 The Institute of Electrical and Electronics Engineers, Inc. Published 2024 by John Wiley & Sons, Inc.

Printed and bound by CPI Group (UK) Ltd, Croydon, CR0 4YY

16/04/2025

14658343-0003